DNA Nanotechnology

Contents

1 DNA nanotechnology   **1**

  1.1 Fundamental concepts   1

     1.1.1 Properties of nucleic acids   1

     1.1.2 Subfields   2

  1.2 Structural DNA nanotechnology   4

     1.2.1 Extended lattices   4

     1.2.2 Discrete structures   5

     1.2.3 Templated assembly   6

  1.3 Dynamic DNA nanotechnology   6

     1.3.1 Nanomechanical devices   6

     1.3.2 Strand displacement cascades   7

  1.4 Applications   7

  1.5 Design   8

     1.5.1 Structural design   8

     1.5.2 Sequence design   9

  1.6 Materials and methods   9

  1.7 History   9

  1.8 See also   10

  1.9 References   10

  1.10 Further reading   15

  1.11 External links   16

2 Coding theory approaches to nucleic acid design   **20**

  2.1 Introduction   20

  2.2 Definitions   21

     2.2.1 Property U   21

     2.2.2 M sequences   22

  2.3 Examples of code construction   22

     2.3.1 1. Code construction using complex Hadamard matrices   22

     2.3.2 2. Code construction via a Binary Mapping   24

2.4 See also 25

2.5 References 25

2.6 External links 25

3 Robert Dirks   **26**

3.1 Early life 26

3.2 Research 27

3.3 Later life and death 27

3.4 Notable works 28

3.5 See also 28

3.6 References 28

4 DNA computing   **30**

4.1 History 30

4.2 Idea 30

4.3 Pros and cons 31

4.4 Examples/Prototypes 31

    4.4.1 Combinatorial problems 31

    4.4.2 Tic Tac Toe game 31

4.5 Capabilities 32

4.6 Methods 32

    4.6.1 DNAzymes 32

    4.6.2 Enzymes 32

    4.6.3 Toehold exchange 33

    4.6.4 Algorithmic self-assembly 33

4.7 Alternative technologies 33

4.8 See also 33

4.9 References 34

4.10 Further reading 35

4.11 External links 36

5 DNA machine   **39**

5.1 References 39

5.2 See also 39

6 DNA origami   **40**

6.1 Overview 40

6.2 Applications 40

6.3 Similar approaches 41

6.4 See also 41

6.5   References 42

7 Inorganic Chromosome Based in Silicon   44

7.1   History 45

7.2   Inchrosil characteristics 45

7.3   Adleman's experiment 46

7.4   Uses for Inchrosil 46

7.5   See also 47

7.6   References 47

7.7   External links 48

8 Thomas LaBean   49

8.1   Works 49

8.2   External links 49

9 List of DNA nanotechnology research groups   50

9.1   North America 50

9.2   Asia 50

9.3   Europe 50

9.4   References 50

10 Nucleic acid design   52

10.1 Fundamental concepts 52

10.2 Approaches 54

     10.2.1 Heuristic methods 54

     10.2.2 Thermodynamic models 55

     10.2.3 Geometrical models 55

10.3 Applications 56

10.4 See also 56

10.5 References 57

10.6 Further reading 58

11 Niles Pierce   59

11.1 Works 59

11.2 Resources 60

11.3 External links 60

12 John Reif   61

12.1 Biography 61

12.2 Research contributions 61

     12.2.1 Research in nanoscience 62

12.3 See also   62

12.4 Publications   62

12.5 References   62

12.6 External links   63

13 Paul W. K. Rothemund   **64**

13.1 Life   64

13.2 References   64

13.3 External links   64

14 Nadrian Seeman   **66**

14.1 Notable publications   66

14.2 See also   67

14.3 References   67

15 Friedrich Simmel   **69**

15.1 Awards and memberships   69

15.2 References   69

15.3 Works   69

15.4 External links   70

16 Transcriptor   **71**

16.1 Background   71

16.2 Invention and description   71

16.3 Impact   71

16.4 References   72

16.5 External links   72

17 Andrew Turberfield   **73**

17.1 References   73

17.2 External links   73

18 Erik Winfree   **74**

18.1 Works   74

18.2 References   74

18.3 External links   75

18.4 Text and image sources, contributors, and licenses   76

    18.4.1 Text   76

    18.4.2 Images   77

    18.4.3 Content license   78

Chapter 1

DNA nanotechnology

DNA nanotechnology is the design and manufacture of artificial nucleic acid structures for technological uses. In this field, nucleic acids are used as non-biological engineering materials for nanotechnology rather than as the carriers of genetic information in living cells. Researchers in the field have created static structures such as two- and three-dimensional crystal lattices, nanotubes, polyhedra, and arbitrary shapes, as well as functional devices such as molecular machines and DNA computers. The field is beginning to be used as a tool to solve basic science problems in structural biology and biophysics, including applications in crystallography and spectroscopy for protein structure determination. Potential applications in molecular scale electronics and nanomedicine are also being investigated.

The conceptual foundation for DNA nanotechnology was first laid out by Nadrian Seeman in the early 1980s, and the field began to attract widespread interest in the mid-2000s. This use of nucleic acids is enabled by their strict base pairing rules, which cause only portions of strands with complementary base sequences to bind together to form strong, rigid double helix structures. This allows for the rational design of base sequences that will selectively assemble to form complex target structures with precisely controlled nanoscale features. A number of assembly methods are used to make these structures, including tile-based structures that assemble from smaller structures, folding structures using the DNA origami method, and dynamically reconfigurable structures using strand displacement techniques. While the field's name specifically references DNA, the same principles have been used with other types of nucleic acids as well, leading to the occasional use of the alternative name **nucleic acid nanotechnology**.

1.1 Fundamental concepts

1.1.1 Properties of nucleic acids

Nanotechnology is often defined as the study of materials and devices with features on a scale below 100 nanometers. DNA nanotechnology, specifically, is an example of bottom-up molecular self-assembly, in which molecular components spontaneously organize into stable structures; the particular form of these structures is induced by the physical and chemical properties of the components selected by the designers.[4] In DNA nanotechnology, the component materials are strands of nucleic acids such as DNA; these strands are often synthetic and are almost always used outside the context of a living cell. DNA is well-suited to nanoscale construction because the binding between two nucleic acid strands depends on simple base pairing rules which are well understood, and form the specific nanoscale structure of the nucleic acid double helix. These qualities make the assembly of nucleic acid structures easy to control through nucleic acid design. This property is absent in other materials used in nanotechnology, including proteins, for which protein design is very difficult, and nanoparticles, which lack the capability for specific assembly on their own.[5]

The structure of a nucleic acid molecule consists of a sequence of nucleotides distinguished by which nucleobase they contain. In DNA, the four bases present are adenine (A), cytosine (C), guanine (G), and thymine (T). Nucleic acids have the property that two molecules will only bind to each other to form a double helix if the two sequences are complementary, meaning that they form matching sequences of base pairs, with A only binding to T, and C only to G.[5][6] Because the

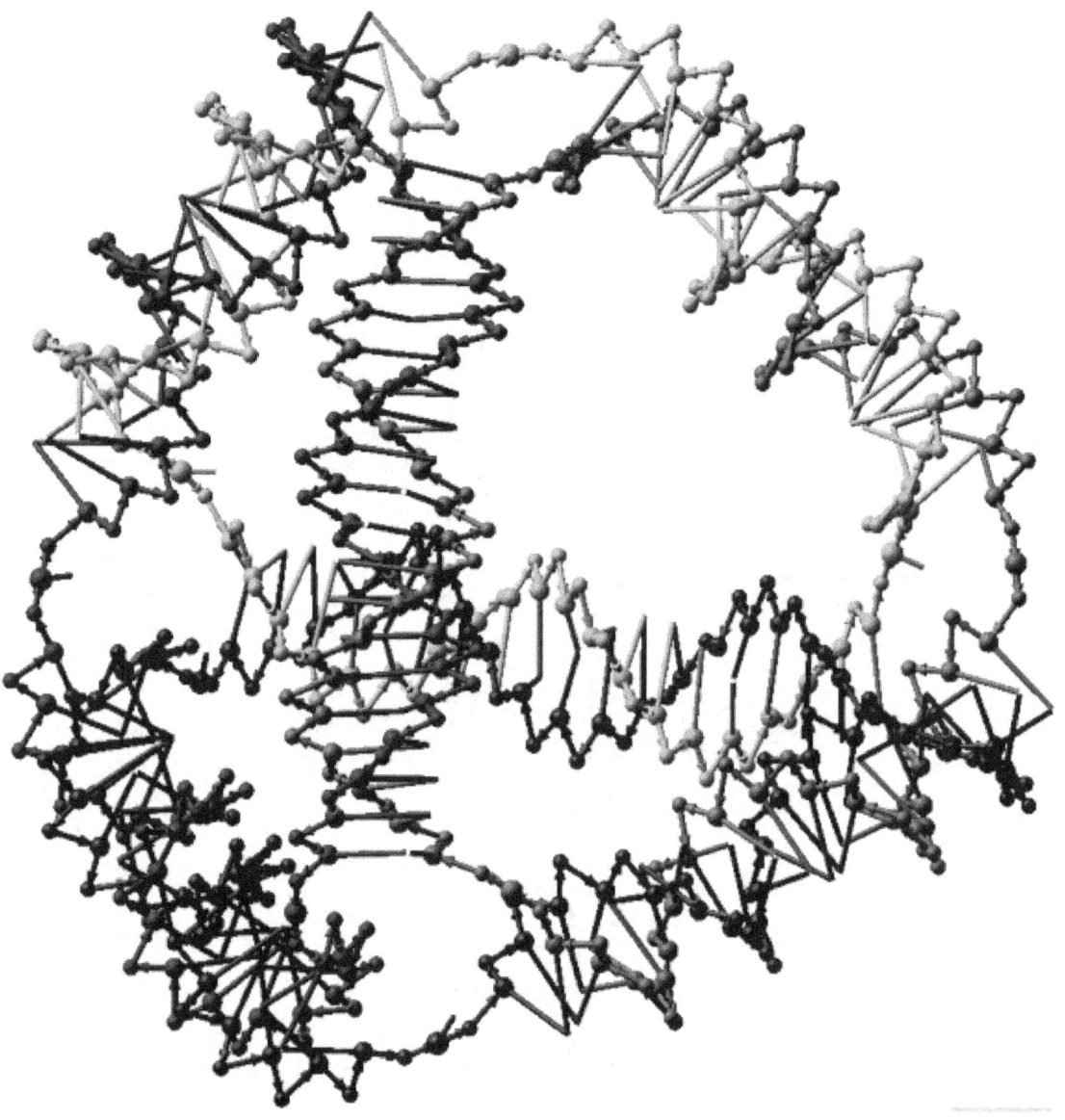

DNA nanotechnology involves the creation of artificial, designed nanostructures out of nucleic acids, such as this DNA tetrahedron.[1] Each edge of the tetrahedron is a 20 base pair DNA double helix, and each vertex is a three-arm junction. The 4 DNA strands that form the 4 tetrahedral faces are color-coded.

formation of correctly matched base pairs is energetically favorable, nucleic acid strands are expected in most cases to bind to each other in the conformation that maximizes the number of correctly paired bases. The sequences of bases in a system of strands thus determine the pattern of binding and the overall structure in an easily controllable way. In DNA nanotechnology, the base sequences of strands are rationally designed by researchers so that the base pairing interactions cause the strands to assemble in the desired conformation.[3][5] While DNA is the dominant material used, structures incorporating other nucleic acids such as RNA and peptide nucleic acid (PNA) have also been constructed.[7][8]

1.1.2   Subfields

DNA nanotechnology is sometimes divided into two overlapping subfields: structural DNA nanotechnology and dynamic DNA nanotechnology. Structural DNA nanotechnology, sometimes abbreviated as SDN, focuses on synthesizing and

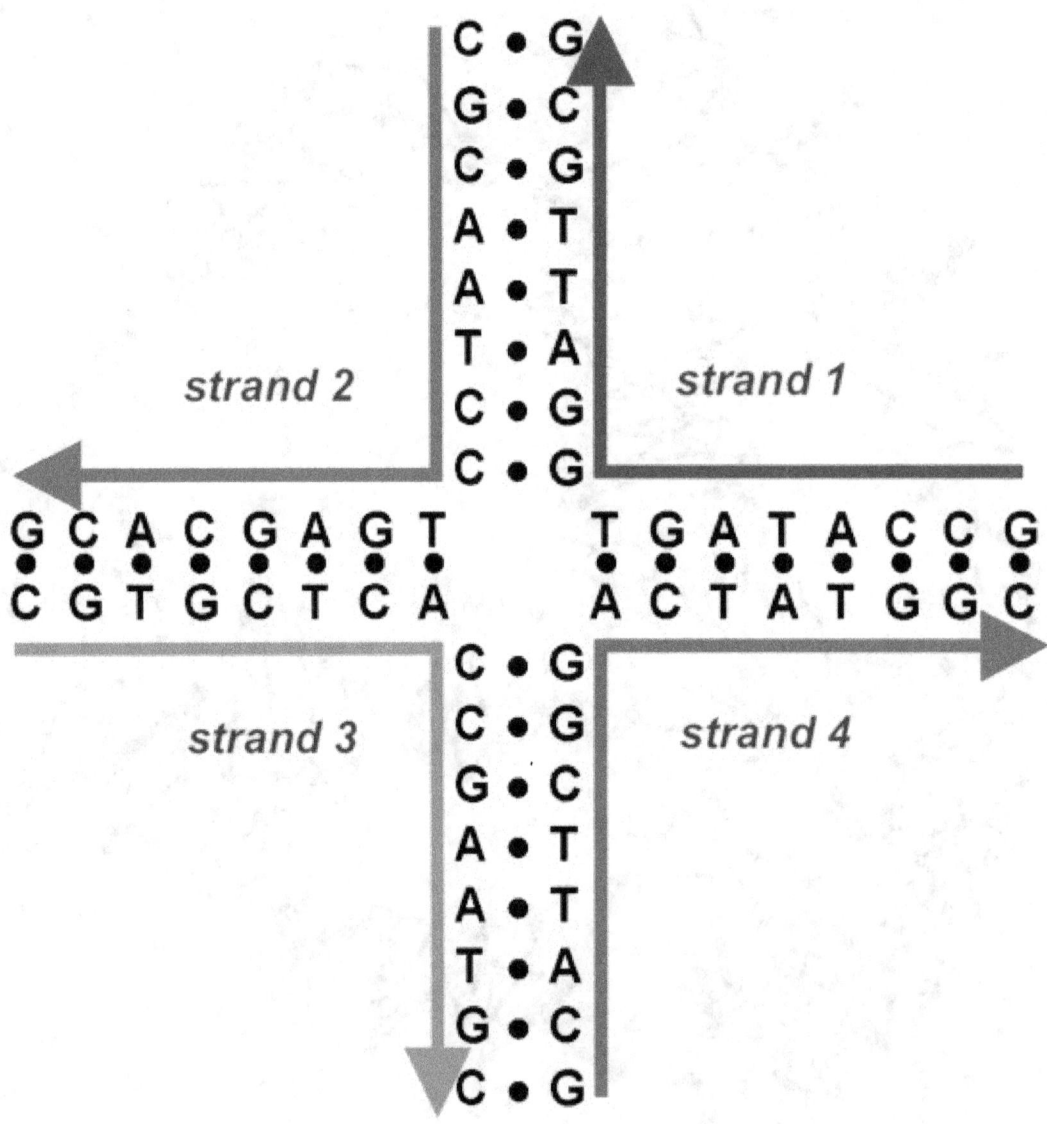

These four strands associate into a DNA four-arm junction because this structure maximizes the number of correct base pairs, with A matched to T and C matched to G.[2][3] See this image for a more realistic model of the four-arm junction showing its tertiary structure.

characterizing nucleic acid complexes and materials that assemble into a static, equilibrium end state. On the other hand, dynamic DNA nanotechnology focuses on complexes with useful non-equilibrium behavior such as the ability to reconfigure based on a chemical or physical stimulus. Some complexes, such as nucleic acid nanomechanical devices, combine features of both the structural and dynamic subfields.[9][10]

The complexes constructed in structural DNA nanotechnology use topologically branched nucleic acid structures containing junctions. (In contrast, most biological DNA exists as an unbranched double helix.) One of the simplest branched structures is a four-arm junction that consists of four individual DNA strands, portions of which are complementary in a specific pattern. Unlike in natural Holliday junctions, each arm in the artificial immobile four-arm junction has a different base sequence, causing the junction point to be fixed at a certain position. Multiple junctions can be combined in the same complex, such as in the widely used double-crossover (DX) motif, which contains two parallel double helical domains with individual strands crossing between the domains at two crossover points. Each crossover point is itself topologically a four-arm junction, but is constrained to a single orientation, as opposed to the flexible single four-arm junction, providing

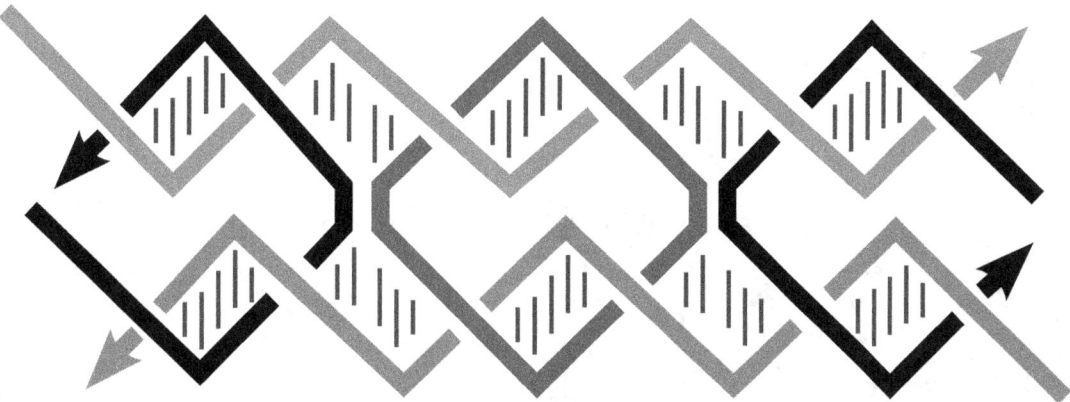

This double-crossover (DX) supramolecular complex consists of five DNA single strands that form two double-helical domains, on the top and the bottom in this image. There are two crossover points where the strands cross from one domain into the other.[2]

a rigidity that makes the DX motif suitable as a structural building block for larger DNA complexes.[3][5]

Dynamic DNA nanotechnology uses a mechanism called toehold-mediated strand displacement to allow the nucleic acid complexes to reconfigure in response to the addition of a new nucleic acid strand. In this reaction, the incoming strand binds to a single-stranded toehold region of a double-stranded complex, and then displaces one of the strands bound in the original complex through a branch migration process. The overall effect is that one of the strands in the complex is replaced with another one.[9] In addition, reconfigurable structures and devices can be made using functional nucleic acids such as deoxyribozymes and ribozymes, which are capable of performing chemical reactions, and aptamers, which can bind to specific proteins or small molecules.[11]

1.2   Structural DNA nanotechnology

Structural DNA nanotechnology, sometimes abbreviated as SDN, focuses on synthesizing and characterizing nucleic acid complexes and materials where the assembly has a static, equilibrium endpoint. The nucleic acid double helix has a robust, defined three-dimensional geometry that makes it possible to predict and design the structures of more complicated nucleic acid complexes. Many such structures have been created, including two- and three-dimensional structures, and periodic, aperiodic, and discrete structures.[10]

1.2.1   Extended lattices

Small nucleic acid complexes can be equipped with sticky ends and combined into larger two-dimensional periodic lattices containing a specific tessellated pattern of the individual molecular tiles.[10] The earliest example of this used double-crossover (DX) complexes as the basic tiles, each containing four sticky ends designed with sequences that caused the DX units to combine into periodic two-dimensional flat sheets that are essentially rigid two-dimensional crystals of DNA.[15][16] Two-dimensional arrays have been made from other motifs as well, including the Holliday junction rhombus lattice,[17] and various DX-based arrays making use of a double-cohesion scheme.[18][19] The top two images at right show examples of tile-based periodic lattices.

Two-dimensional arrays can be made to exhibit aperiodic structures whose assembly implements a specific algorithm, exhibiting one form of DNA computing.[20] The DX tiles can have their sticky end sequences chosen so that they act as Wang tiles, allowing them to perform computation. A DX array whose assembly encodes an XOR operation has been demonstrated; this allows the DNA array to implement a cellular automaton that generates a fractal known as the Sierpinski gasket. The third image at right shows this type of array.[14] Another system has the function of a binary counter, displaying a representation of increasing binary numbers as it grows. These results show that computation can

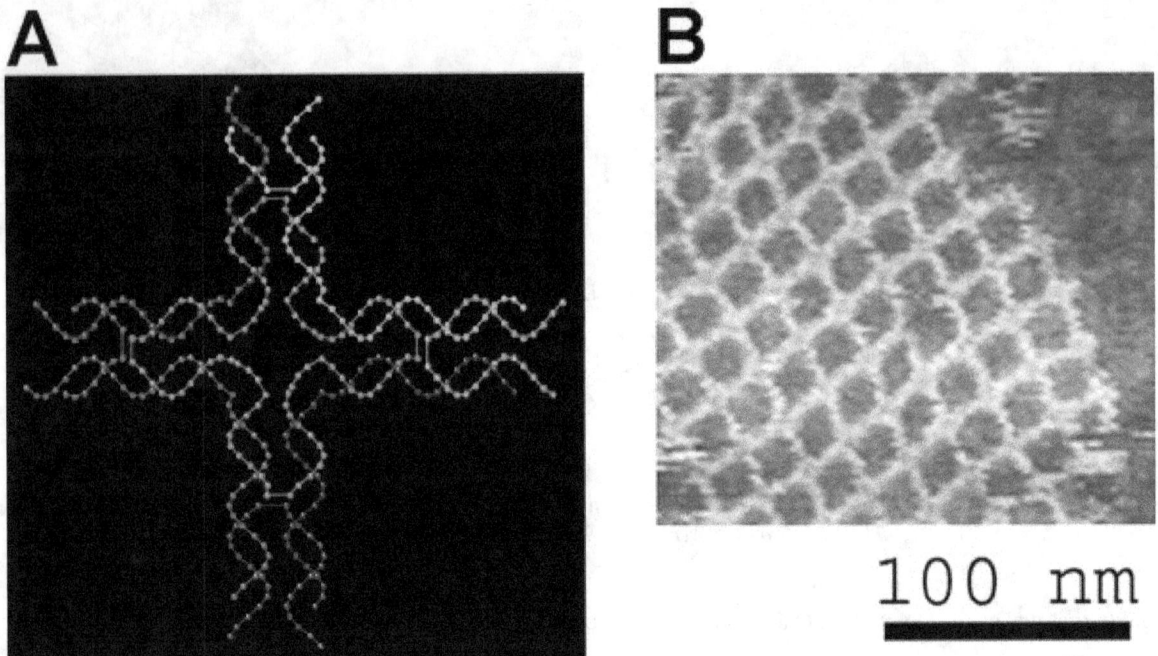

Left, *a model of a DNA tile used to make another two-dimensional periodic lattice.* Right, *an atomic force micrograph of the assembled lattice.*[12][13]

be incorporated into the assembly of DNA arrays.[21]

DX arrays have been made to form hollow nanotubes 4–20 nm in diameter, essentially two-dimensional lattices which curve back upon themselves.[22] These DNA nanotubes are somewhat similar in size and shape to carbon nanotubes, and while they lack the electrical conductance of carbon nanotubes, DNA nanotubes are more easily modified and connected to other structures. One of many schemes for constructing DNA nanotubes uses a lattice of curved DX tiles that curls around itself and closes into a tube.[23] In an alternative method that allows the circumference to be specified in a simple, modular fashion using single-stranded tiles, the rigidity of the tube is an emergent property.[24]

The creation of three-dimensional lattices out of DNA was the earliest goal of DNA nanotechnology, but this proved to be one of the most difficult to realize. Success using a motif based on the concept of tensegrity, a balance between tension and compression forces, was finally reported in 2009.[20][25]

1.2.2 Discrete structures

Researchers have synthesized a number of three-dimensional DNA complexes that each have the connectivity of a polyhedron, such as a cube or octahedron, meaning that the DNA duplexes trace the edges of a polyhedron with a DNA junction at each vertex.[26] The earliest demonstrations of DNA polyhedra were very work-intensive, requiring multiple ligations and solid-phase synthesis steps to create catenated polyhedra.[27] Subsequent work yielded polyhedra whose synthesis was much easier. These include a DNA octahedron made from a long single strand designed to fold into the correct conformation,[28] and a tetrahedron that can be produced from four DNA strands in a single step, pictured at the top of this article.[1]

Nanostructures of arbitrary, non-regular shapes are usually made using the DNA origami method. These structures consist of a long, natural virus strand as a "scaffold", which is made to fold into the desired shape by computationally designed short "staple" strands. This method has the advantages of being easy to design, as the base sequence is predetermined by the scaffold strand sequence, and not requiring high strand purity and accurate stoichiometry, as most other DNA nanotechnology methods do. DNA origami was first demonstrated for two-dimensional shapes, such as a smiley face and a coarse map of the Western Hemisphere.[26][29] Solid three-dimensional structures can be made by using parallel DNA helices arranged in a honeycomb pattern,[30] and structures with two-dimensional faces can be made to fold into a hollow overall three-dimensional shape, akin to a cardboard box. These can be programmed to open and reveal or release a molecular cargo in response to a stimulus, making them potentially useful as programmable molecular cages.[31][32]

1.2.3 Templated assembly

Nucleic acid structures can be made to incorporate molecules other than nucleic acids, sometimes called heteroelements, including proteins, metallic nanoparticles, quantum dots, and fullerenes. This allows the construction of materials and devices with a range of functionalities much greater than is possible with nucleic acids alone. The goal is to use the self-assembly of the nucleic acid structures to template the assembly of the nanoparticles hosted on them, controlling their position and in some cases orientation.[26][33] Many of these schemes use a covalent attachment scheme, using oligonucleotides with amide or thiol functional groups as a chemical handle to bind the heteroelements. This covalent binding scheme has been used to arrange gold nanoparticles on a DX-based array,[34] and to arrange streptavidin protein molecules into specific patterns on a DX array.[35] A non-covalent hosting scheme using Dervan polyamides on a DX array was used to arrange streptavidin proteins in a specific pattern on a DX array.[36] Carbon nanotubes have been hosted on DNA arrays in a pattern allowing the assembly to act as a molecular electronic device, a carbon nanotube field-effect transistor.[37] In addition, there are nucleic acid metallization methods, in which the nucleic acid is replaced by a metal which assumes the general shape of the original nucleic acid structure,[38] and schemes for using nucleic acid nanostructures as lithography masks, transferring their pattern into a solid surface.[39]

1.3 Dynamic DNA nanotechnology

Dynamic DNA nanotechnology focuses on creating nucleic acid systems with designed dynamic functionalities related to their overall structures, such as computation and mechanical motion. There is some overlap between structural and dynamic DNA nanotechnology, as structures can be formed through annealing and then reconfigured dynamically, or can be made to form dynamically in the first place.[26][40]

1.3.1 Nanomechanical devices

Main article: DNA machine

DNA complexes have been made that change their conformation upon some stimulus, making them one form of nanorobotics. These structures are initially formed in the same way as the static structures made in structural DNA nanotechnology, but are designed so that dynamic reconfiguration is possible after the initial assembly.[9][40] The earliest such device made use of the transition between the B-DNA and Z-DNA forms to respond to a change in buffer conditions by undergoing a twisting motion.[41] This reliance on buffer conditions, however, caused all devices to change state at the same time. Subsequent systems could change states based upon the presence of control strands, allowing multiple devices to be independently operated in solution. Some examples of such systems are a "molecular tweezers" design that has an open and a closed state,[42] a device that could switch from a paranemic-crossover (PX) conformation to a double-junction (JX2) conformation, undergoing rotational motion in the process,[43] and a two-dimensional array that could dynamically expand and contract in response to control strands.[44] Structures have also been made that dynamically open or close, potentially acting as a molecular cage to release or reveal a functional cargo upon opening.[31][45][46]

DNA walkers are a class of nucleic acid nanomachines that exhibit directional motion along a linear track. A large number of schemes have been demonstrated.[40] One strategy is to control the motion of the walker along the track using control strands that need to be manually added in sequence.[47][48] Another approach is to make use of restriction enzymes or deoxyribozymes to cleave the strands and cause the walker to move forward, which has the advantage of running autonomously.[49][50] A later system could walk upon a two-dimensional surface rather than a linear track, and demonstrated the ability to selectively pick up and move molecular cargo.[51] Additionally, a linear walker has been demonstrated that performs DNA-templated synthesis as the walker advances along the track, allowing autonomous multistep chemical synthesis directed by the walker.[52] The synthetic DNA walkers' function is similar to that of the proteins dynein and kinesin.[53]

1.3.2 Strand displacement cascades

Cascades of strand displacement reactions can be used for either computational or structural purposes. An individual strand displacement reaction involves revealing a new sequence in response to the presence of some initiator strand. Many such reactions can be linked into a cascade where the newly revealed output sequence of one reaction can initiate another strand displacement reaction elsewhere. This in turn allows for the construction of chemical reaction networks with many components, exhibiting complex computational and information processing abilities. These cascades are made energetically favorable through the formation of new base pairs, and the entropy gain from disassembly reactions. Strand displacement cascades allow for isothermal operation of the assembly or computational process, as opposed to traditional nucleic acid assembly's requirement for a thermal annealing step, where the temperature is raised and then slowly lowered to ensure proper formation of the desired structure. They can also support catalytic functionality of the initiator species, where less than one equivalent of the initiator can cause the reaction to go to completion.[9][54]

Strand displacement complexes can be used to make molecular logic gates capable of complex computation.[55] Unlike traditional electronic computers, which use electric current as inputs and outputs, molecular computers use the concentrations of specific chemical species as signals. In the case of nucleic acid strand displacement circuits, the signal is the presence of nucleic acid strands that are released or consumed by binding and unbinding events to other strands in displacement complexes. This approach has been used to make logic gates such as AND, OR, and NOT gates.[56] More recently, a four-bit circuit was demonstrated that can compute the square root of the integers 0–15, using a system of gates containing 130 DNA strands.[57]

Another use of strand displacement cascades is to make dynamically assembled structures. These use a hairpin structure for the reactants, so that when the input strand binds, the newly revealed sequence is on the same molecule rather than disassembling. This allows new opened hairpins to be added to a growing complex. This approach has been used to make simple structures such as three- and four-arm junctions and dendrimers.[54]

1.4 Applications

DNA nanotechnology provides one of the few ways to form designed, complex structures with precise control over nanoscale features. The field is beginning to see application to solve basic science problems in structural biology and biophysics. The earliest such application envisaged for the field, and one still in development, is in crystallography, where molecules that are difficult to crystallize in isolation could be arranged within a three-dimensional nucleic acid lattice, allowing determination of their structure. Another application is the use of DNA origami rods to replace liquid crystals in residual dipolar coupling experiments in protein NMR spectroscopy; using DNA origami is advantageous because, unlike liquid crystals, they are tolerant of the detergents needed to suspend membrane proteins in solution. DNA walkers have been used as nanoscale assembly lines to move nanoparticles and direct chemical synthesis. Furthermore, DNA origami structures have aided in the biophysical studies of enzyme function and protein folding.[10][58]

DNA nanotechnology is moving towards potential real-world applications. The ability of nucleic acid arrays to arrange other molecules indicates its potential applications in molecular scale electronics. The assembly of a nucleic acid structure could be used to template the assembly of a molecular electronic elements such as molecular wires, providing a method for nanometer-scale control of the placement and overall architecture of the device analogous to a molecular breadboard.[10][26] DNA nanotechnology has been compared to the concept of programmable matter because of the coupling of computation to its material properties.[59]

In a study conducted by a group of scientists from iNANO center and CDNA Center in Aarhus university (Aarhus), researchers were able to construct a small multi-switchable 3D DNA Box Origami. The proposed nanoparticle was characterized by AFM, TEM and FRET. The constructed box was shown to have a unique reclosing mechanism, which enabled it to repeatedly open and close in response to a unique set of DNA or RNA keys. The authors proposed that this "DNA device can potentially be used for a broad range of applications such as controlling the function of single molecules, controlled drug delivery, and molecular computing.".[60]

There are potential applications for DNA nanotechnology in nanomedicine, making use of its ability to perform computation in a biocompatible format to make "smart drugs" for targeted drug delivery. One such system being investigated uses a hollow DNA box containing proteins that induce apoptosis, or cell death, that will only open when in proximity to a cancer cell.[58][61] There has additionally been interest in expressing these artificial structures in engineered living bacterial cells, most likely using the transcribed RNA for the assembly, although it is unknown whether these complex structures are able to efficiently fold or assemble in the cell's cytoplasm. If successful, this could enable directed evolution of nucleic acid nanostructures.[26] Scientists at Oxford University reported the self-assembly of four short strands of synthetic DNA into a cage which is capable of entering cells and surviving for at least 48 hours. The fluorescently labeled DNA tetrahedra were found to remain intact in the laboratory cultured human kidney cells despite the attack by cellular enzymes after two days. This experiment showed the potential of drug delivery inside the living cells using the DNA 'cage'.[62][63] A DNA tetrahedron was used to deliver RNA Interference (RNAi) in a mouse model, reported a team of researchers in MIT. Delivery of the interfering RNA for treatment has showed some success using polymer or lipid, but there are limitations of safety and imprecise targeting, in addition to short shelf life in the blood stream. The DNA nanostructure created by the team consists of six strands of DNA to form a tetrahedron, with a single strand of RNA affixed to each of the six edges. The tetrahedron is further equipped with targeting protein, three folate molecules, which lead the DNA nanoparticles to the abundant folate receptors found on some tumors. The result showed that the gene expression targeted by the RNAi, luciferase, dropped by more than half. This study shows promise in using DNA nanotechnology as an effective tool to deliver treatment using the emerging RNA Interference technology.[64][65]

1.5 Design

DNA nanostructures must be rationally designed so that the individual nucleic acid strands will assemble into the desired structures. This process usually begins with the specification of a desired target structure or functionality. Then, the overall secondary structure of the target complex is determined, specifying the arrangement of nucleic acid strands within the structure, and which portions of those strands should be bound to each other. The last step is the primary structure design, which is the specification of the actual base sequences of each nucleic acid strand.[22][66]

1.5.1 Structural design

The first step in designing a nucleic acid nanostructure is to decide how a given structure should be represented by a specific arrangement of nucleic acid strands. This design step determines the secondary structure, or the positions of the base pairs that hold the individual strands together in the desired shape.[22] Several approaches have been demonstrated:

- **Tile-based structures.** This approach breaks the target structure into smaller units with strong binding between the strands contained in each unit, and weaker interactions between the units. It is often used to make periodic lattices, but can also be used to implement algorithmic self-assembly, making them a platform for DNA computing. This was the dominant design strategy used from the mid-1990s until the mid-2000s, when the DNA origami methodology was developed.[22][67]

- **Folding structures.** An alternative to the tile-based approach, folding approaches make the nanostructure from a single long strand. This long strand can either have a designed sequence that folds due to its interactions with itself, or it can be folded into the desired shape by using shorter, "staple" strands. This latter method is called DNA origami, which allows the creation of nanoscale two- and three-dimensional shapes (see Discrete structures below).[26][29]

- **Dynamic assembly.** This approach directly controls the kinetics of DNA self-assembly, specifying all of the intermediate steps in the reaction mechanism in addition to the final product. This is done using starting materials which adopt a hairpin structure; these then assemble into the final conformation in a cascade reaction, in a specific order (see Strand displacement cascades below). This approach has the advantage of proceeding isothermally, at a constant temperature. This is in contrast to the thermodynamic approaches, which require a thermal annealing step where a temperature change is required to trigger the assembly and favor proper formation of the desired structure.[26][54]

1.5.2 Sequence design

Main article: Nucleic acid design

After any of the above approaches are used to design the secondary structure of a target complex, an actual sequence of nucleotides that will form into the desired structure must be devised. Nucleic acid design is the process of assigning a specific nucleic acid base sequence to each of a structure's constituent strands so that they will associate into a desired conformation. Most methods have the goal of designing sequences so that the target structure has the lowest energy, and is thus the most thermodynamically favorable, while incorrectly assembled structures have higher energies and are thus disfavored. This is done either through simple, faster heuristic methods such as sequence symmetry minimization, or by using a full nearest-neighbor thermodynamic model, which is more accurate but slower and more computationally intensive. Geometric models are used to examine tertiary structure of the nanostructures and to ensure that the complexes are not overly strained.[66][68]

Nucleic acid design has similar goals to protein design. In both, the sequence of monomers is designed to favor the desired target structure and to disfavor other structures. Nucleic acid design has the advantage of being much computationally easier than protein design, because the simple base pairing rules are sufficient to predict a structure's energetic favorability, and detailed information about the overall three-dimensional folding of the structure is not required. This allows the use of simple heuristic methods that yield experimentally robust designs. However, nucleic acid structures are less versatile than proteins in their functionality because of proteins' increased ability to fold into complex structures, as well as the limited chemical diversity of the four nucleotides as compared to the twenty proteinogenic amino acids.[68]

1.6 Materials and methods

The sequences of the DNA strands making up a target structure are designed computationally, using molecular modeling and thermodynamic modeling software.[66][68] The nucleic acids themselves are then synthesized using standard oligonucleotide synthesis methods, usually automated in an oligonucleotide synthesizer, and strands of custom sequences are commercially available.[69] Strands can be purified by denaturing gel electrophoresis if needed,[70] and precise concentrations determined via any of several nucleic acid quantitation methods using ultraviolet absorbance spectroscopy.[71]

The fully formed target structures can be verified using native gel electrophoresis, which gives size and shape information for the nucleic acid complexes. An electrophoretic mobility shift assay can assess whether a structure incorporates all desired strands.[72] Fluorescent labeling and Förster resonance energy transfer (FRET) are sometimes used to characterize the structure of the complexes.[73]

Nucleic acid structures can be directly imaged by atomic force microscopy, which is well suited to extended two-dimensional structures, but less useful for discrete three-dimensional structures because of the microscope tip's interaction with the fragile nucleic acid structure; transmission electron microscopy and cryo-electron microscopy are often used in this case. Extended three-dimensional lattices are analyzed by X-ray crystallography.[74][75]

1.7 History

The conceptual foundation for DNA nanotechnology was first laid out by Nadrian Seeman in the early 1980s.[76] Seeman's original motivation was to create a three-dimensional DNA lattice for orienting other large molecules, which would simplify their crystallographic study by eliminating the difficult process of obtaining pure crystals. This idea had reportedly come to him in late 1980, after realizing the similarity between the woodcut *Depth* by M. C. Escher and an array of DNA six-arm junctions.[3][77] A number of natural branched DNA structures were known at the time, including the DNA replication fork and the mobile Holliday junction, but Seeman's insight was that immobile nucleic acid junctions could be created by properly designing the strand sequences to remove symmetry in the assembled molecule, and that these immobile junctions could in principle be combined into rigid crystalline lattices. The first theoretical paper proposing this scheme was published in 1982, and the first experimental demonstration of an immobile DNA junction was published the following year.[5][26]

In 1991, Seeman's laboratory published a report on the synthesis of a cube made of DNA, the first synthetic three-dimensional nucleic acid nanostructure, for which he received the 1995 Feynman Prize in Nanotechnology. This was followed by a DNA truncated octahedron. However, it soon became clear that these structures, polygonal shapes with flexible junctions as their vertices, were not rigid enough to form extended three-dimensional lattices. Seeman developed the more rigid double-crossover (DX) motif, and in 1998, in collaboration with Erik Winfree, published the creation of two-dimensional lattices of DX tiles.[3][76][78] These tile-based structures had the advantage that they provided the capability to implement DNA computing, which was demonstrated by Winfree and Paul Rothemund in their 2004 paper on the algorithmic self-assembly of a Sierpinski gasket structure, and for which they shared the 2006 Feynman Prize in Nanotechnology. Winfree's key insight was that the DX tiles could be used as Wang tiles, meaning that their assembly was capable of performing computation.[76] The synthesis of a three-dimensional lattice was finally published by Seeman in 2009, nearly thirty years after he had set out to achieve it.[58]

New capabilities continued to be discovered for designed DNA structures throughout the 2000s. The first DNA nanomachine—a motif that changes its structure in response to an input—was demonstrated in 1999 by Seeman. An improved system, which was the first nucleic acid device to make use of toehold-mediated strand displacement, was demonstrated by Bernard Yurke the following year. The next advance was to translate this into mechanical motion, and in 2004 and 2005, a number of DNA walker systems were demonstrated by the groups of Seeman, Niles Pierce, Andrew Turberfield, and Chengde Mao.[40] The idea of using DNA arrays to template the assembly of other molecules such as nanoparticles and proteins, first suggested by Bruche Robinson and Seeman in 1987,[79] was demonstrated in 2002 by Seeman, Kiehl et al.[80] and subsequently by numerous other groups.

In 2006, Rothemund first demonstrated the DNA origami technique for easily and robustly creating folded DNA structures of arbitrary shape. Rothemund had conceived of this method as being conceptually intermediate between Seeman's DX lattices, which used many short strands, and William Shih's DNA octahedron, which consisted mostly of one very long strand. Rothemund's DNA origami contains a long strand whose folding is assisted by a number of short strands. This method allowed the creation of much larger structures than were previously possible, and which are less technically demanding to design and synthesize.[78] DNA origami was the cover story of *Nature* on March 15, 2006.[29] Rothemund's research demonstrating two-dimensional DNA origami structures was followed by the demonstration of solid three-dimensional DNA origami by Douglas *et al.* in 2009,[30] while the labs of Jørgen Kjems and Yan demonstrated hollow three-dimensional structures made out of two-dimensional faces.[58]

DNA nanotechnology was initially met with some skepticism due to the unusual non-biological use of nucleic acids as materials for building structures and doing computation, and the preponderance of proof of principle experiments that extended the capabilities of the field but were far from actual applications. Seeman's 1991 paper on the synthesis of the DNA cube was rejected by the journal *Science* after one reviewer praised its originality while another criticized it for its lack of biological relevance. By the early 2010s, however, the field was considered to have increased its capabilities to the point that applications for basic science research were beginning to be realized, and practical applications in medicine and other fields were beginning to be considered feasible.[58][81] The field had grown from very few active laboratories in 2001 to at least 60 in 2010, which increased the talent pool and thus the number of scientific advances in the field during that decade.[20]

1.8   See also

- International Society for Nanoscale Science, Computation, and Engineering

- Nanobiotechnology

- Molecular models of DNA

- List of nucleic acid simulation software

1.9   References

[1] **DNA polyhedra:** Goodman, Russel P.; Schaap, Iwan A. T.; Tardin, C. F.; Erben, Christof M.; Berry, Richard M.; Schmidt, C.F.; Turberfield, Andrew J. (9 December 2005). "Rapid chiral assembly of rigid DNA building blocks for molecular nanofab-

rication". *Science* **310** (5754): 1661–1665. Bibcode:2005Sci...310.1661G. doi:10.1126/science.1120367. PMID 16339440.

[2] **Overview:** Mao, Chengde (December 2004). "The emergence of complexity: lessons from DNA". *PLoS Biology* **2** (12): 2036–2038. doi:10.1371/journal.pbio.0020431. PMC 535573. PMID 15597116.

[3] **Overview:** Seeman, Nadrian C. (June 2004). "Nanotechnology and the double helix". *Scientific American* **290** (6): 64–75. doi:10.1038/scientificamerican0604-64. PMID 15195395.

[4] **Background:** Pelesko, John A. (2007). *Self-assembly: the science of things that put themselves together*. New York: Chapman & Hall/CRC. pp. 5, 7. ISBN 978-1-58488-687-7.

[5] **Overview:** Seeman, Nadrian C. (2010). "Nanomaterials based on DNA". *Annual Review of Biochemistry* **79**: 65–87. doi:10.1146/annurev-biochem-060308-102244. PMC 3454582. PMID 20222824.

[6] **Background:** Long, Eric C. (1996). "Fundamentals of nucleic acids". In Hecht, Sidney M. *Bioorganic chemistry: nucleic acids*. New York: Oxford University Press. pp. 4–10. ISBN 0-19-508467-5.

[7] **RNA nanotechnology:** Chworos, Arkadiusz; Severcan, Isil; Koyfman, Alexey Y.; Weinkam, Patrick; Oroudjev, Emin; Hansma, Helen G.; Jaeger, Luc (2004). "Building Programmable Jigsaw Puzzles with RNA". *Science* **306** (5704): 2068–2072. Bibcode:2004Sci...306.2 doi:10.1126/science.1104686. PMID 15604402.

[8] **RNA nanotechnology:** Guo, Peixuan (2010). "The Emerging Field of RNA Nanotechnology". *Nature Nanotechnology* **5** (12): 833–842. Bibcode:2010NatNa...5..833G. doi:10.1038/nnano.2010.231. PMC 3149862. PMID 21102465.

[9] **Dynamic DNA nanotechnology:** Zhang, D. Y.; Seelig, G. (February 2011). "Dynamic DNA nanotechnology using strand-displacement reactions". *Nature Chemistry* **3** (2): 103–113. Bibcode:2011NatCh...3..103Z. doi:10.1038/nchem.957. PMID 21258382.

[10] **Structural DNA nanotechnology:** Seeman, Nadrian C. (November 2007). "An overview of structural DNA nanotechnology". *Molecular Biotechnology* **37** (3): 246–257. doi:10.1007/s12033-007-0059-4. PMC 3479651. PMID 17952671.

[11] **Dynamic DNA nanotechnology:** Lu, Y.; Liu, J. (December 2006). "Functional DNA nanotechnology: Emerging applications of DNAzymes and aptamers". *Current Opinion in Biotechnology* **17** (6): 580–588. doi:10.1016/j.copbio.2006.10.004. PMID 17056247.

[12] **Other arrays:** Strong, Michael (March 2004). "Protein Nanomachines". *PLoS Biology* **2** (3): e73. doi:10.1371/journal.pbio.0020073. PMC 368168. PMID 15024422.

[13] Yan, H.; Park, S. H.; Finkelstein, G.; Reif, J. H.; Labean, T. H. (26 September 2003). "DNA-templated self-assembly of protein arrays and highly conductive nanowires". *Science* **301** (5641): 1882–1884. Bibcode:2003Sci...301.1882Y. doi:10.1126/science.1089389. PMID 14512621.

[14] **Algorithmic self-assembly:** Rothemund, Paul W. K.; Papadakis, Nick; Winfree, Erik (December 2004). "Algorithmic self-assembly of DNA Sierpinski triangles". *PLoS Biology* **2** (12): 2041–2053. doi:10.1371/journal.pbio.0020424. PMC 534809. PMID 15583715.

[15] **DX arrays:** Winfree, Erik; Liu, Furong; Wenzler, Lisa A.; Seeman, Nadrian C. (6 August 1998). "Design and self-assembly of two-dimensional DNA crystals". *Nature* **394** (6693): 529–544. Bibcode:1998Natur.394..539W. doi:10.1038/28998. PMID 9707114.

[16] **DX arrays:** Liu, Furong; Sha, Ruojie; Seeman, Nadrian C. (10 February 1999). "Modifying the surface features of two-dimensional DNA crystals". *Journal of the American Chemical Society* **121** (5): 917–922. doi:10.1021/ja982824a.

[17] **Other arrays:** Mao, Chengde; Sun, Weiqiong; Seeman, Nadrian C. (16 June 1999). "Designed two-dimensional DNA Holliday junction arrays visualized by atomic force microscopy". *Journal of the American Chemical Society* **121** (23): 5437–5443. doi:10.1021/ja9900398.

[18] **Other arrays:** Constantinou, Pamela E.; Wang, Tong; Kopatsch, Jens; Israel, Lisa B.; Zhang, Xiaoping; Ding, Baoquan; Sherman, William B.; Wang, Xing; Zheng, Jianping; Sha, Ruojie; Seeman, Nadrian C. (21 September 2006). "Double cohesion in structural DNA nanotechnology". *Organic and Biomolecular Chemistry* **4** (18): 3414–3419. doi:10.1039/b605212f. PMC 3491902. PMID 17036134.

[19] **Other arrays:** Mathieu, Frederick; Liao, Shiping; Kopatsch, Jens; Wang, Tong; Mao, Chengde; Seeman, Nadrian C. (April 2005). "Six-helix bundles designed from DNA". *Nano Letters* **5** (4): 661–665. Bibcode:2005NanoL...5..661M. doi:10.1021/nl050084f. PMC 3464188. PMID 15826105.

[20] **History**: Seeman, Nadrian (9 June 2010). "Structural DNA nanotechnology: growing along with Nano Letters". *Nano Letters* **10** (6): 1971–1978. Bibcode:2010NanoL..10.1971S. doi:10.1021/nl101262u. PMC 2901229. PMID 20486672.

[21] **Algorithmic self-assembly:** Barish, Robert D.; Rothemund, Paul W. K.; Winfree, Erik (December 2005). "Two computational primitives for algorithmic self-assembly: copying and counting". *Nano Letters* **5** (12): 2586–2592. Bibcode:2005NanoL...5.2586B. doi:10.1021/nl052038l. PMID 16351220.

[22] **Design:** Feldkamp, U.; Niemeyer, C. M. (13 March 2006). "Rational design of DNA nanoarchitectures". *Angewandte Chemie International Edition* **45** (12): 1856–1876. doi:10.1002/anie.200502358. PMID 16470892.

[23] **DNA nanotubes:** Rothemund, Paul W. K.; Ekani-Nkodo, Axel; Papadakis, Nick; Kumar, Ashish; Fygenson, Deborah Kuchnir & Winfree, Erik (22 December 2004). "Design and Characterization of Programmable DNA Nanotubes". *Journal of the American Chemical Society* **126** (50): 16344–16352. doi:10.1021/ja0443191. PMID 15600335.

[24] **DNA nanotubes:** Yin, P.; Hariadi, R. F.; Sahu, S.; Choi, H. M. T.; Park, S. H.; Labean, T. H.; Reif, J. H. (8 August 2008). "Programming DNA Tube Circumferences". *Science* **321** (5890): 824–826. Bibcode:2008Sci...321..824Y. doi:10.1126/science.1157312. PMID 18687961.

[25] **Three-dimensional arrays:** Zheng, Jianping; Birktoft, Jens J.; Chen, Yi; Wang, Tong; Sha, Ruojie; Constantinou, Pamela E.; Ginell, Stephan L.; Mao, Chengde; Seeman, Nadrian C. (3 September 2009). "From molecular to macroscopic via the rational design of a self-assembled 3D DNA crystal". *Nature* **461** (7260): 74–77. Bibcode:2009Natur.461...74Z. doi:10.1038/nature08274. PMC 2764300. PMID 19727196.

[26] **Overview:** Pinheiro, A. V.; Han, D.; Shih, W. M.; Yan, H. (December 2011). "Challenges and opportunities for structural DNA nanotechnology". *Nature Nanotechnology* **6** (12): 763–772. Bibcode:2011NatNa...6..763P. doi:10.1038/nnano.2011.187. PMC 3334823. PMID 22056726.

[27] **DNA polyhedra:** Zhang, Yuwen; Seeman, Nadrian C. (1 March 1994). "Construction of a DNA-truncated octahedron". *Journal of the American Chemical Society* **116** (5): 1661–1669. doi:10.1021/ja00084a006.

[28] **DNA polyhedra:** Shih, William M.; Quispe, Joel D.; Joyce, Gerald F. (12 February 2004). "A 1.7-kilobase single-stranded DNA that folds into a nanoscale octahedron". *Nature* **427** (6975): 618–621. Bibcode:2004Natur.427..618S. doi:10.1038/nature02307. PMID 14961116.

[29] **DNA origami:** Rothemund, Paul W. K. (16 March 2006). "Folding DNA to create nanoscale shapes and patterns". *Nature* **440** (7082): 297–302. Bibcode:2006Natur.440..297R. doi:10.1038/nature04586. PMID 16541064.

[30] **DNA origami:** Douglas, Shawn M.; Dietz, Hendrik; Liedl, Tim; Högberg, Björn; Graf, Franziska; Shih, William M. (21 May 2009). "Self-assembly of DNA into nanoscale three-dimensional shapes". *Nature* **459** (7245): 414–418. Bibcode:2009Natur.459..414D. doi:10.1038/nature08016. PMC 2688462. PMID 19458720.

[31] **DNA boxes:** Andersen, Ebbe S.; Dong, Mingdong; Nielsen, Morten M.; Jahn, Kasper; Subramani, Ramesh; Mamdouh, Wael; Golas, Monika M.; Sander, Bjoern; et al. (7 May 2009). "Self-assembly of a nanoscale DNA box with a controllable lid". *Nature* **459** (7243): 73–76. Bibcode:2009Natur.459...73A. doi:10.1038/nature07971. PMID 19424153.

[32] **DNA boxes:** Ke, Yonggang; Sharma, Jaswinder; Liu, Minghui; Jahn, Kasper; Liu, Yan; Yan, Hao (10 June 2009). "Scaffolded DNA origami of a DNA tetrahedron molecular container". *Nano Letters* **9** (6): 2445–2447. Bibcode:2009NanoL...9.2445K. doi:10.1021/nl901165f. PMID 19419184.

[33] **Overview:** Endo, M.; Sugiyama, H. (12 October 2009). "Chemical approaches to DNA nanotechnology". *ChemBioChem* **10** (15): 2420–2443. doi:10.1002/cbic.200900286. PMID 19714700.

[34] **Nanoarchitecture:** Zheng, Jiwen; Constantinou, Pamela E.; Micheel, Christine; Alivisatos, A. Paul; Kiehl, Richard A.; Seeman Nadrian C. (July 2006). "2D Nanoparticle Arrays Show the Organizational Power of Robust DNA Motifs". *Nano Letters* **6** (7): 1502–1504. Bibcode:2006NanoL...6.1502Z. doi:10.1021/nl060994c. PMC 3465979. PMID 16834438.

[35] **Nanoarchitecture:** Park, Sung Ha; Pistol, Constantin; Ahn, Sang Jung; Reif, John H.; Lebeck, Alvin R.; Dwyer, Chris; LaBean, Thomas H. (October 2006). "Finite-size, fully addressable DNA tile lattices formed by hierarchical assembly procedures". *Angewandte Chemie* **118** (40): 749–753. doi:10.1002/ange.200690141.

[36] **Nanoarchitecture:** Cohen, Justin D.; Sadowski, John P.; Dervan, Peter B. (22 October 2007). "Addressing single molecules on DNA nanostructures". *Angewandte Chemie International Edition* **46** (42): 7956–7959. doi:10.1002/anie.200702767. PMID 17763481.

[37] **Nanoarchitecture:** Maune, Hareem T.; Han, Si-Ping; Barish, Robert D.; Bockrath, Marc; Goddard III, William A.; Rothemund, Paul W. K.; Winfree, Erik (January 2009). "Self-assembly of carbon nanotubes into two-dimensional geometries using DNA origami templates". *Nature Nanotechnology* **5** (1): 61–66. Bibcode:2010NatNa...5...61M. doi:10.1038/nnano.2009.311. PMID 19898497.

[38] **Nanoarchitecture:** Liu, J.; Geng, Y.; Pound, E.; Gyawali, S.; Ashton, J. R.; Hickey, J.; Woolley, A. T.; Harb, J. N. (22 March 2011). "Metallization of branched DNA origami for nanoelectronic circuit fabrication". *ACS Nano* **5** (3): 2240–2247. doi:10.1021/nn1035075. PMID 21323323.

[39] **Nanoarchitecture:** Deng, Z.; Mao, C. (6 August 2004). "Molecular lithography with DNA nanostructures". *Angewandte Chemie International Edition* **43** (31): 4068. doi:10.1002/anie.200460257.

[40] **DNA machines:** Bath, Jonathan; Turberfield, Andrew J. (May 2007). "DNA nanomachines". *Nature Nanotechnology* **2** (5): 275–284. Bibcode:2007NatNa...2..275B. doi:10.1038/nnano.2007.104. PMID 18654284.

[41] **DNA machines:** Mao, Chengde; Sun, Weiqiong; Shen, Zhiyong; Seeman, Nadrian C. (14 January 1999). "A DNA nanomechanical device based on the B-Z transition". *Nature* **397** (6715): 144–146. Bibcode:1999Natur.397..144M. doi:10.1038/16437. PMID 9923675.

[42] **DNA machines:** Yurke, Bernard; Turberfield, Andrew J.; Mills, Allen P., Jr; Simmel, Friedrich C.; Neumann, Jennifer L. (10 August 2000). "A DNA-fuelled molecular machine made of DNA". *Nature* **406** (6796): 605–609. Bibcode:2000Natur.406..605Y. doi:10.1038/35020524. PMID 10949296.

[43] **DNA machines:** Yan, Hao; Zhang, Xiaoping; Shen, Zhiyong; Seeman, Nadrian C. (3 January 2002). "A robust DNA mechanical device controlled by hybridization topology". *Nature* **415** (6867): 62–65. Bibcode:2002Natur.415...62Y. doi:10.1038/415062a. PMID 11780115.

[44] **DNA machines:** Feng, L.; Park, S. H.; Reif, J. H.; Yan, H. (22 September 2003). "A two-state DNA lattice switched by DNA nanoactuator". *Angewandte Chemie* **115** (36): 4478. doi:10.1002/ange.200351818.

[45] **DNA machines:** Goodman, R. P.; Heilemann, M.; Doose, S. R.; Erben, C. M.; Kapanidis, A. N.; Turberfield, A. J. (February 2008). "Reconfigurable, braced, three-dimensional DNA nanostructures". *Nature Nanotechnology* **3** (2): 93–96. Bibcode:2008NatNa...3...93 doi:10.1038/nnano.2008.3. PMID 18654468.

[46] **Applications:** Douglas, Shawn M.; Bachelet, Ido; Church, George M. (17 February 2012). "A logic-gated nanorobot for targeted transport of molecular payloads". *Science* **335** (6070): 831–834. Bibcode:2012Sci...335..831D. doi:10.1126/science.1214081.

[47] **DNA walkers:** Shin, Jong-Shik; Pierce, Niles A. (8 September 2004). "A synthetic DNA walker for molecular transport". *Journal of the American Chemical Society* **126** (35): 10834–10835. doi:10.1021/ja047543j. PMID 15339155.

[48] **DNA walkers:** Sherman, William B.; Seeman, Nadrian C. (July 2004). "A precisely controlled DNA biped walking device". *Nano Letters* **4** (7): 1203–1207. Bibcode:2004NanoL...4.1203S. doi:10.1021/nl049527q.

[49] **DNA walkers:** Tian, Ye; He, Yu; Chen, Yi; Yin, Peng; Mao, Chengde (11 July 2005). "A DNAzyme that walks processively and autonomously along a one-dimensional track". *Angewandte Chemie* **117** (28): 4429–4432. doi:10.1002/ange.200500703.

[50] **DNA walkers:** Bath, Jonathan; Green, Simon J.; Turberfield, Andrew J. (11 July 2005). "A free-running DNA motor powered by a nicking enzyme". *Angewandte Chemie International Edition* **44** (28): 4358–4361. doi:10.1002/anie.200501262.

[51] **Functional DNA walkers:** Lund, Kyle; Manzo, Anthony J.; Dabby, Nadine; Michelotti, Nicole; Johnson-Buck, Alexander; Nangreave, Jeanette; Taylor, Steven; Pei, Renjun; Stojanovic, Milan N.; Walter, Nils G.; Winfree, Erik; Yan, Hao (13 May 2010). "Molecular robots guided by prescriptive landscapes". *Nature* **465** (7295): 206–210. Bibcode:2010Natur.465..206L. doi:10.1038/nature09012. PMC 2907518. PMID 20463735.

[52] **Functional DNA walkers:** He, Yu; Liu, David R. (November 2010). "Autonomous multistep organic synthesis in a single isothermal solution mediated by a DNA walker". *Nature Nanotechnology* **5** (11): 778–782. Bibcode:2010NatNa...5..778H. doi:10.1038/nnano.2010.190. PMC 2974042. PMID 20935654.

[53] "Recent progress on DNA based walkers". *www.sciencedirect.com.proxy1.lib.uwo.ca*. Retrieved 2015-09-28.

[54] **Kinetic assembly:** Yin, Peng; Choi, Harry M. T.; Calvert, Colby R.; Pierce, Niles A. (17 January 2008). "Programming biomolecular self-assembly pathways". *Nature* **451** (7176): 318–322. Bibcode:2008Natur.451..318Y. doi:10.1038/nature06451. PMID 18202654.

[55] **Fuzzy and Boolean logic gates based on DNA:** Zadegan, R. M.; Jepsen, M. D. E.; Hildebrandt, L. L.; Birkedal, V.; Kjems, J. R. (2015). "Construction of a Fuzzy and Boolean Logic Gates Based on DNA". *Small* **11** (15): 1811. doi:10.1002/smll.201402755. PMID 25565140.

[56] **Strand displacement cascades:** Seelig, G.; Soloveichik, D.; Zhang, D. Y.; Winfree, E. (8 December 2006). "Enzyme-free nucleic acid logic circuits". *Science* **314** (5805): 1585–1588. Bibcode:2006Sci...314.1585S. doi:10.1126/science.1132493. PMID 17158324.

[57] **Strand displacement cascades:** Qian, Lulu; Winfree, Erik (3 June 2011). "Scaling up digital circuit computation with DNA strand displacement cascades". *Science* **332** (6034): 1196–1201. Bibcode:2011Sci...332.1196Q. doi:10.1126/science.1200520. PMID 21636773.

[58] **History/applications:** Service, Robert F. (3 June 2011). "DNA nanotechnology grows up". *Science* **332** (6034): 1140–1143. doi:10.1126/science.332.6034.1140.

[59] **Applications:** Rietman, Edward A. (2001). *Molecular engineering of nanosystems.* Springer. pp. 209–212. ISBN 978-0-387-98988-4. Retrieved 17 April 2011.

[60] M. Zadegan, Reza; et. al. (2012). "Construction of a 4 Zeptoliters Switchable 3D DNA Box Origami". *ACS Nano* **6** (11): 10050–10053. doi:10.1021/nn303767b.

[61] **Applications:** Jungmann, Ralf; Renner, Stephan; Simmel, Friedrich C. (March 2008). "From DNA nanotechnology to synthetic biology". *HFSP journal* **2** (2): 99–109. doi:10.2976/1.2896331. PMC 2645571. PMID 19404476.

[62] Lovy, Howard (5 July 2011). "DNA cages can unleash meds inside cells". fiercedrugdelivery.com. Retrieved 22 September 2013.

[63] Walsh, Anthony; Yin, Hai; Erben, Christoph; Wood, Matthew; Turberfield, Andrew (2011). "DNA Cage Delivery to Mammalian Cells". *ACS Nano* (ACS Publications) **5** (7): 5427–5432. doi:10.1021/nn2005574. PMID 21696187.

[64] Trafton, Anne (4 June 2012). "Researchers achieve RNA interference, in a lighter package". MIT News. Retrieved 22 September 2013.

[65] Lee, Hyukjin; Lytton-Jean, Abigail; Chen, Yi; Love, Kevin; Park, Angela; Karagiannis, Emmanouil; Sehgal, Alfica; Querbes, William; et al. (2012). "Molecularly self-assembled nucleic acid nanoparticles for targeted in vivo siRNA delivery" (PDF). *Nature Nanotechnology* (Nature) **7** (6): 389–393. Bibcode:2012NatNa...7..389L. doi:10.1038/NNANO.2012.73.

[66] **Design:** Brenneman, Arwen; Condon, Anne (25 September 2002). "Strand design for biomolecular computation". *Theoretical Computer Science* **287**: 39–58. doi:10.1016/S0304-3975(02)00135-4.

[67] **Overview:** Lin, Chenxiang; Liu, Yan; Rinker, Sherri; Yan, Hao (11 August 2006). "DNA tile based self-assembly: building complex nanoarchitectures". *ChemPhysChem* **7** (8): 1641–1647. doi:10.1002/cphc.200600260. PMID 16832805.

[68] **Design:** Dirks, Robert M.; Lin, Milo; Winfree, Erik; Pierce, Niles A. (15 February 2004). "Paradigms for computational nucleic acid design". *Nucleic Acids Research* **32** (4): 1392–1403. doi:10.1093/nar/gkh291. PMC 390280. PMID 14990744.

[69] **Methods:** Ellington, A.; Pollard, J. D. (1 May 2001). "Synthesis and purification of oligonucleotides". *Current Protocols in Molecular Biology.* doi:10.1002/0471142727.mb0211s42. ISBN 0471142727.

[70] **Methods:** Ellington, A.; Pollard, J. D. (1 May 2001). "Purification of oligonucleotides using denaturing polyacrylamide gel electrophoresis". *Current Protocols in Molecular Biology.* doi:10.1002/0471142727.mb0212s42. ISBN 0471142727.

[71] **Methods:** Gallagher, S. R.; Desjardins, P. (1 July 2011). "Quantitation of nucleic acids and proteins". *Current Protocols Essential Laboratory Techniques.* doi:10.1002/9780470089941.et0202s5. ISBN 0470089938.

[72] **Methods:** Chory, J.; Pollard, J. D. (1 May 2001). "Separation of small DNA fragments by conventional gel electrophoresis". *Current Protocols in Molecular Biology.* doi:10.1002/0471142727.mb0207s47. ISBN 0471142727.

[73] **Methods:** Walter, N. G. (1 February 2003). "Probing RNA structural dynamics and function by fluorescence resonance energy transfer (FRET)". *Current Protocols in Nucleic Acid Chemistry.* doi:10.1002/0471142700.nc1110s11. ISBN 0471142700.

[74] **Methods:** Lin, C.; Ke, Y.; Chhabra, R.; Sharma, J.; Liu, Y.; Yan, H. (2011). "Synthesis and Characterization of Self-Assembled DNA Nanostructures". In Zuccheri, G. and Samorì, B. *DNA Nanotechnology: Methods and Protocols.* Methods in Molecular Biology **749**. pp. 1–11. doi:10.1007/978-1-61779-142-0_1. ISBN 978-1-61779-141-3.

[75] **Methods:** Bloomfield, Victor A.; Crothers, Donald M.; Tinoco, Jr., Ignacio (2000). *Nucleic acids: structures, properties, and functions*. Sausalito, Calif: University Science Books. pp. 84–86, 396–407. ISBN 0-935702-49-0.

[76] **History:** Pelesko, John A. (2007). *Self-assembly: the science of things that put themselves together*. New York: Chapman & Hall/CRC. pp. 201, 242, 259. ISBN 978-1-58488-687-7.

[77] **History:** See "Current crystallization protocol". Nadrian Seeman Lab. for a statement of the problem, and "DNA cages containing oriented guests". Nadrian Seeman Laboratory. for the proposed solution.

[78] **DNA origami:** Rothemund, Paul W. K. (2006). "Scaffolded DNA origami: from generalized multicrossovers to polygonal networks". In Chen, Junghuei; Jonoska, Natasha; Rozenberg, Grzegorz. *Nanotechnology: science and computation*. Natural Computing Series. New York: Springer. pp. 3–21. doi:10.1007/3-540-30296-4_1. ISBN 978-3-540-30295-7.

[79] **Nanoarchitecture:** Robinson, Bruche H.; Seeman, Nadrian C. (August 1987). "The design of a biochip: a self-assembling molecular-scale memory device". *Protein Engineering* **1** (4): 295–300. doi:10.1093/protein/1.4.295. PMID 3508280.

[80] **Nanoarchitecture:** Xiao, Shoujun; Liu, Furong; Rosen, Abbey E.; Hainfeld, James F.; Seeman, Nadrian C.; Musier-Forsyth, Karin; Kiehl, Richard A. (August 2002). "Selfassembly of metallic nanoparticle arrays by DNA scaffolding". *Journal of Nanoparticle Research* **4** (4): 313–317. doi:10.1023/A:1021145208328.

[81] **History:** Hopkin, Karen (August 2011). "Profile: 3-D seer". *The Scientist*. Retrieved 8 August 2011.

1.10   Further reading

General:

- Seeman, Nadrian C. (June 2004). "Nanotechnology and the double helix". *Scientific American* **290** (6): 64–75. doi:10.1038/scientificamerican0604-64. PMID 15195395.—An article written for laypeople by the founder of the field

- Seeman, Nadrian C. (9 June 2010). "Structural DNA nanotechnology: growing along with Nano Letters". *Nano Letters* **10** (6): 1971–1978. Bibcode:2010NanoL..10.1971S. doi:10.1021/nl101262u. PMC 2901229. PMID 20486672.—A review of results in the period 2001–2010

- Seeman, Nadrian C. (2010). "Nanomaterials based on DNA". *Annual Review of Biochemistry* **79**: 65–87. doi:10.1146/annurev-biochem-060308-102244. PMC 3454582. PMID 20222824.—A more comprehensive review including both old and new results in the field

- Service, Robert F. (3 June 2011). "DNA nanotechnology grows up". *Science* **332** (6034): 1140–1143. doi:10.1126/science.332. and doi:10.1126/science.332.6034.1142.—A news article focusing on the history of the field and development of new applications

- Zadegan, Reza M.; Norton, Michael L. (June 2012). "Structural DNA Nanotechnology: From Design to Applications". *Int. J. Mol. Sci.* **13** (6): 7149–7162. doi:10.3390/ijms13067149. PMC 3397516. PMID 22837684.—A very recent and comprehensive review in the field

Specific subfields:

- Bath, Jonathan; Turberfield, Andrew J. (5 May 2007). "DNA nanomachines". *Nature Nanotechnology* **2** (5): 275–284. Bibcode:2007NatNa...2..275B. doi:10.1038/nnano.2007.104. PMID 18654284.—A review of nucleic acid nanomechanical devices

- Feldkamp, Udo; Niemeyer, Christof M. (13 March 2006). "Rational design of DNA nanoarchitectures". *Angewandte Chemie International Edition* **45** (12): 1856–76. doi:10.1002/anie.200502358. PMID 16470892.—A review coming from the viewpoint of secondary structure design

- Lin, Chenxiang; Liu, Yan; Rinker, Sherri; Yan, Hao (11 August 2006). "DNA tile based self-assembly: building complex nanoarchitectures". *ChemPhysChem* **7** (8): 1641–1647. doi:10.1002/cphc.200600260. PMID 16832805.—A minireview specifically focusing on tile-based assembly

- Zhang, David Yu; Seelig, Georg (February 2011). "Dynamic DNA nanotechnology using strand-displacement reactions". *Nature Chemistry* **3** (2): 103–113. Bibcode:2011NatCh...3..103Z. doi:10.1038/nchem.957. PMID 21258382.—A review of DNA systems making use of strand displacement mechanisms

1.11   External links

- International Society for Nanoscale Science, Computation and Engineering

- What is Bionanotechnology?—a video introduction to DNA nanotechnology

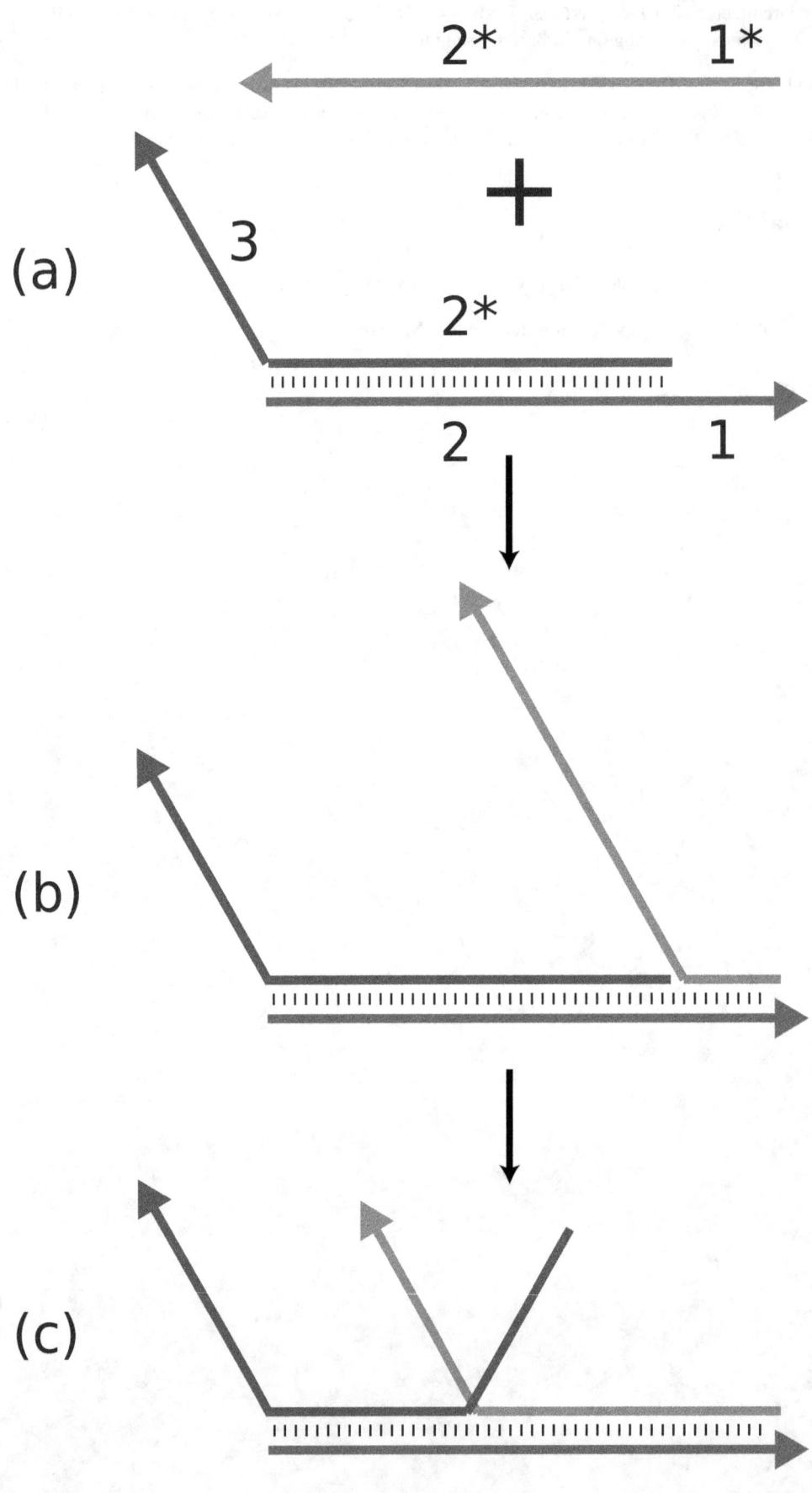

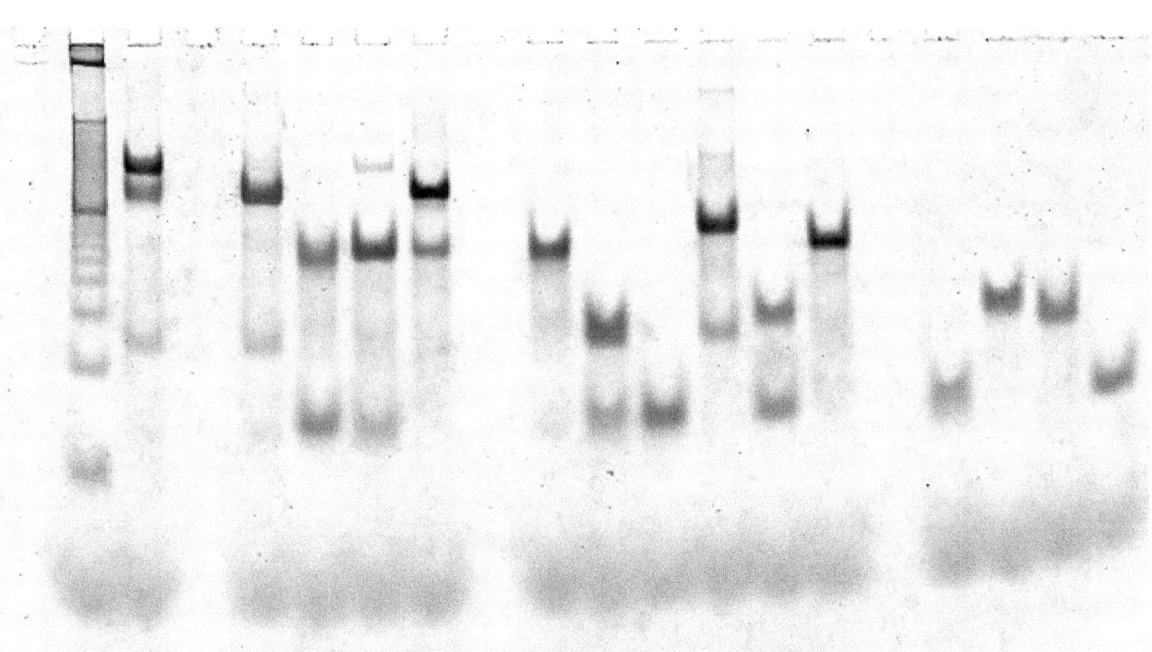

Gel electrophoresis methods, such as this formation assay on a DX complex, are used to ascertain whether the desired structures are forming properly. Each vertical lane contains a series of bands, where each band is characteristic of a particular reaction intermediate.

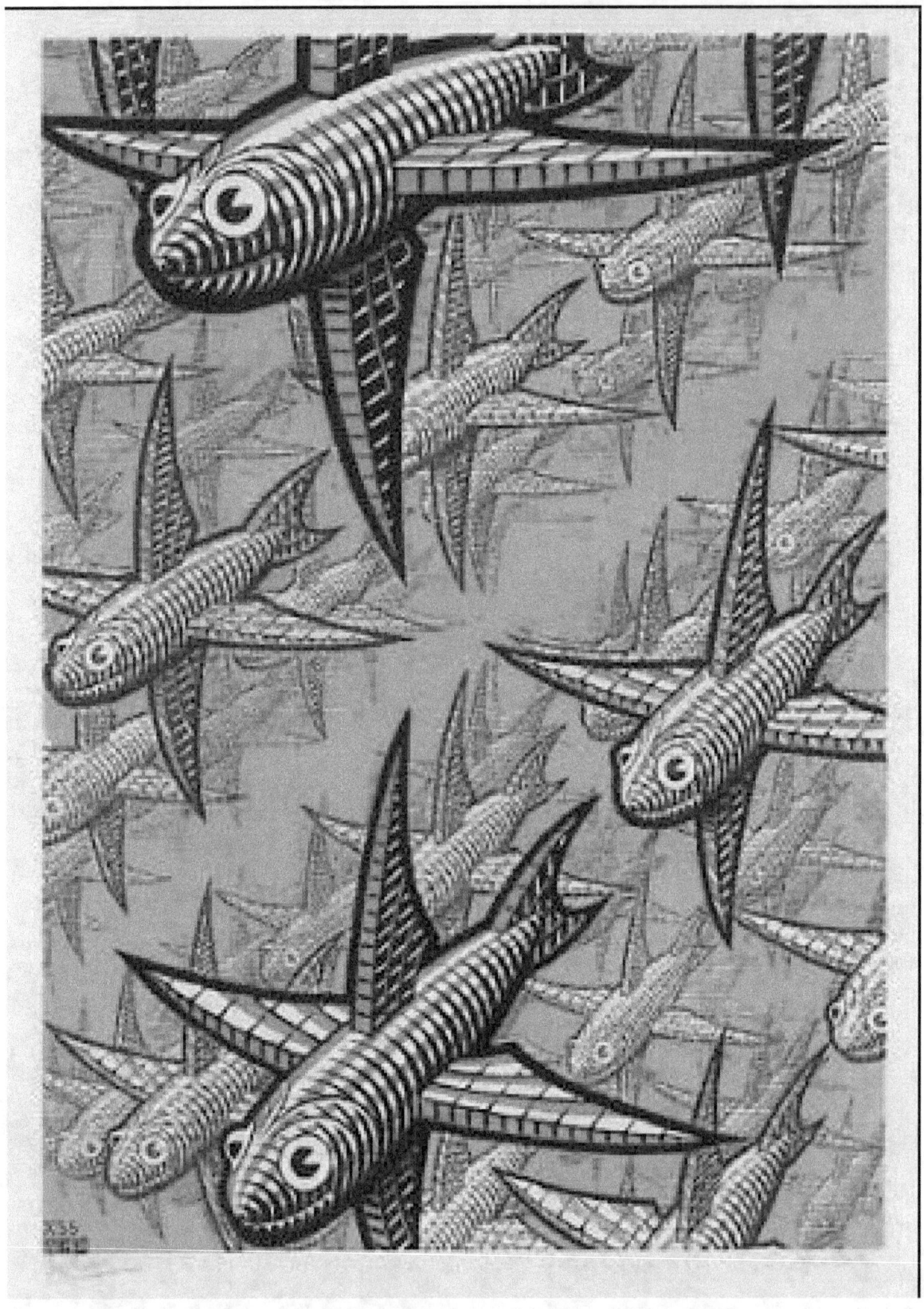

The woodcut Depth *(pictured) by M. C. Escher reportedly inspired Nadrian Seeman to consider using three-dimensional lattices of DNA to orient hard-to-crystallize molecules. This led to the beginning of the field of DNA nanotechnology.*

Chapter 2

Coding theory approaches to nucleic acid design

DNA code construction refers to the application of coding theory to the design of nucleic acid systems for the field of DNA–based computation.

2.1   Introduction

DNA sequences are known to appear in the form of double helices in living cells, in which one DNA strand is hybridized to its complementary strand through a series of hydrogen bonds. For the purpose of this entry, we shall focus on only oligonucleotides. DNA computing involves allowing synthetic oligonucleotide strands to hybridize in such a way as to perform computation. DNA computing requires that the self-assembly of the oligonucleotide strands happen in such a way that hybridization should occur in a manner compatible with the goals of computation.

The field of DNA computing was established in Leonard M. Adelman's seminal paper.[1] His work is significant for a number of reasons:

- It shows how one could use the highly parallel nature of computation performed by DNA to solve problems that are difficult or almost impossible to solve using the traditional methods.

- It's an example of computation at a molecular level, on the lines of nanocomputing, and this potentially is a major advantage as far as the information density on storage media is considered, which can never be reached by the semiconductor industry.

- It demonstrates unique aspects of DNA as a data structure.

This capability for massively parallel computation in DNA computing can be exploited in solving many computational problems on an enormously large scale such as cell-based computational systems for cancer diagnostics and treatment, and ultra-high density storage media.[2]

This selection of codewords (sequences of DNA oligonucleotides) is a major hurdle in itself due to the phenomenon of secondary structure formation (in which DNA strands tend to fold onto themselves during hybridization and hence rendering themselves useless in further computations. This is also known as self-hybridization). The Nussinov-Jacobson[3] algorithm is used to predict secondary structures and also to identify certain design criteria that reduce the possibility of secondary structure formation in a codeword. In essence this algorithm shows how the presence of a cyclic structure in a DNA code reduces the complexity of the problem of testing the codewords for secondary structures.

Novel constructions of such codes include using cyclic reversible extended Goppa codes, generalized Hadamard matrices, and a binary approach. Before diving into these constructions, we shall revisit certain fundamental genetic terminology.

20

The motivation for the theorems presented in this article, is that they concur with the Nussinov - Jacobson algorithm, in that the existence of cyclic structure helps in reducing complexity and thus prevents secondary structure formation. i.e. these algorithms satisfy some or all the design requirements for DNA oligonucleotides at the time of hybridization (which is the core of the DNA computing process) and hence do not suffer from the problems of self - hybridization.

2.2 Definitions

A DNA code is simply a set of sequences over the alphabet $\mathcal{Q} = \{A, T, C, G\}$.

Each purine base is the Watson-Crick complement of a unique pyrimidine base (and vice versa) – adenine and thymine form a complementary pair, as do guanine and cytosine. This pairing can be described as follows – $\bar{A} = T, \bar{T} = A, \bar{C} = G, \bar{G} = C$.

Such pairing is chemically very stable and strong. However, pairing of mismatching bases does occur at times due to biological mutations.

Most of the focus on DNA coding has been on constructing large sets of DNA codewords with prescribed minimum distance properties. For this purpose let us lay down the required groundwork to proceed further.

Let $q = q_1 q_2 .... q_n$ be a word of length n over the alphabet $\mathcal{Q}$. For $1 \leqslant i \leqslant j \leqslant n$, we will use the notation $q_{[i,j]}$ to denote the subsequence $q_i q_{i+1} ... q_j$. Furthermore, the sequence obtained by reversing q will be denoted as q^R . The *Watson-Crick complement*, or the reverse-complement of q, is defined to be $q^{RC} = \bar{q}_n \bar{q}_{n-1} ... \bar{q}_1$, where $\bar{q}_i$ denotes the *Watson-Crick complement* base pair of q_i .

For any pair of length- n words p and q over $\mathcal{Q}$, the Hamming distance $d_H(p,q)$ is the number of positions i at which $p_i \neq q_i$. Further, define *reverse-Hamming distance* as $d_H{}^R(p,q) = d_H(p, q^R)$. Similarly, *reverse-complement Hamming distance* is $d_H^{RC}(p,q) = d_H(p, q^{RC})$. (where RC stands for *reverse complement*)

Another important code design consideration linked to the process of oligonucleotide hybridization pertains to the GC content of sequences in a DNA code. The *GC-content*, $w_{GC}(q)$, of a DNA sequence $q = q_1 q_2 .... q_n$ is defined to be the number of indices i such that $q_i \in \{G, C\}$. A DNA code in which all codewords have the same GC-content, w , is called a constant *GC-content code*.

A *generalized Hadamard matrix* $H \equiv H(n, \mathbb{C}_m)$ is an $n \times n$ square matrix with entries taken from the set of m th roots of unity, $\mathbb{C}_m = \{e^{-2\pi i l}/m , l = 0, ..., m - 1\}$, that satisfies $HH^* = nI$. Here I denotes the identity matrix of order n , while * stands for complex-congugation. We will only concern ourselves with the case $m = p$ for some prime p . A necessary condition for the existence of generalized Hadamard matrices $H(n, \mathbb{C}_p)$ is that $p|n$. The *exponent matrix*, $E(n, \mathbb{Z}_p)$, of $H(n, \mathbb{C}_p)$ is the $n \times n$ matrix with the entries in $Z_p = \{0, 1, 2, ..., p - 1\}$, is obtained by replacing each entry $(e^{-2\pi i l})$ in $H(n, \mathbb{C}_p)$ by the exponent l .

The elements of the Hadamard exponent matrix lie in the Galois field $GF(p)$, and its row vectors constitute the codewords of what shall be called a generalized Hadamard code.

Here, the elements of E lie in the Galois field $GF(p)$.

By definition, a generalized Hadamard matrix H in its standard form has only 1s in its first row and column. The $(n - 1) \times (n - 1)$ square matrix formed by the remaining entries of H is called the *core* of H , and the corresponding submatrix of the exponent matrix E is called the *core* of construction. Thus, by omission of the all-zero first column cyclic generalized Hadamard codes are possible, whose codewords are the row vectors of the punctured matrix.

Also, the rows of such an exponent matrix satisfy the following two properties: (i) in each of the nonzero rows of the exponent matrix, each element of $\mathbb{Z}_p$ appears a constant number, n/p , of times; and (ii) the Hamming distance between any two rows is $n(p - 1)/p$.[4]

2.2.1 Property U

Let $C_p = 1, x, x2, ..., xp - I$ be the cyclic group generated by x , where $x = \exp(2\pi j/p)$ is a complex primitive p th root of unity, and $p > 2$ is a fixed prime. Further, let $A = (x^{a_i})$, $B = (x^{b_i})$ denote arbitrary vectors over C_p

which are of length $N = pt$, where t is a positive integer. Define the collection of differences between exponents $Q = a_i - b_i \mod p : i = 1, 2, ..., N$, where n_q is the multiplicity of element q of $GF(p)$ which appears in Q .[4]

Vector Q is said to satisfy Property U iff each element q of $GF(p)$ appears in Q exactly t times ($n_q = t, q = 0, 1, ..., p-1$)

The following lemma is of fundamental importance in constructing generalized Hadamard codes.

Lemma. Orthogonality of vectors over C_p - For fixed primes p , arbitrary vectors A, B of length $N = pt$, whose elements are from C_p , are orthogonal if the vector Q satisfies *Property U*, where Q is the collection of differences   mod p between the Hadamard exponents associated with A, B .

2.2.2   M sequences

Let V be an arbitrary vector of length N whose elements are in the finite field $GF(p)$, where p is a prime. Let the elements of vector V constitute the first period of an infinite sequence $a(V)$ which is periodic of period N . If N is the smallest period for conceiving a subsequence, the sequence is called an M-sequence, or a sequence of maximal least period obtained by cycling N elements. If, when the elements of the ordered set V are permuted arbitrarily to yield V^* , the sequence $a(V^*)$ is an M-sequence, the sequence $a(V)$ is called **M-invariant**. The theorems that follow present conditions that ensure invariance in an *M sequence*. In conjunction with a certain uniformity property of polynomial coeffecients, these conditions yield a simple method by which complex Hadamard matrices with cyclic core can be constructed.

The goal as outlined at the head of this article is to find cyclic matrix $E = E_c$ whose elements are in Galois field $GF(p)$ and whose dimension is $N = p^n - 1$. The rows of E will be the nonzero codewords of a linear cyclic code K , if and only if there is polynomial $g(x)$ with coefficients in $GF(p)$, which is a proper divisor of $x^N - 1$ and which generates K . In order to have N nonzero codewords, $g(x)$ must be of degree $N - n$. Further, in order to generate a cyclic Hadamard core, the vector (of coefficients of) $g(x)$ when operated upon with the cyclic shift operation must be of period N , and the vector difference of two arbitrary rows of E (augmented with zero) must satisfy the uniformity condition of Butson,[5] previously referred to as *Property U*. One necessary condition for N -periodicity is that $x^N - 1 = g(x)h(x)$, where $h(x)$ is monic irreducible over.[6] The approach here is to replace the last requirement with the condition that the coefficients of the vector $[0, g(x)]$ be uniformly distributed over $GF(p)$, each residue $0, 1, ..., p - 1$ appears the same number of times (Property U). This heuristic approach has succeeded for all cases tried, and a proof that it always produces a cyclic core is given below.

2.3   Examples of code construction

2.3.1   1. Code construction using complex Hadamard matrices

Construction algorithm

Consider all monic irreducible polynomials $h(x)$ over $GF(p)$ which are of degree n , and which permit a suitable companion $g(x)$ of degree $N - n$ such that $g(x)h(x) = x^N - 1$, where also vector $[0, g(x)]$ satisfies *Property U*. This requires only a simple computer algorithm for long division over $GF(p)$. Since $h(x)|x^N - 1$, the ideal generated by $g(x)$,   mod $x^N - 1$, will be a cyclic code K . Moreover, *Property U* guarantees the nonzero codewords form a cyclic matrix, each row being of period N under cyclic permutation, which serves as a cyclic core for Hadamard matrix $H(p, pn)$. As an example, a cyclic core for $H(3, 9)$ results from the companions $h(x) = x^2 + x + 2$ and $g(x) = x^6 + 2x^5 + 2x^4 + 2x^2 + x + 1$. The coefficients of g indicate that $0, 1, 6$ is the relative difference set,   mod 8 .

Theorem

Let p be a prime and $N + 1 = pn$, with $g(x)$ a monic polynomial of degree $N - n$ whose extended vector of coefficients $C = [c_0, c_1, ..., c_{N-1}]$ are elements of $GF(p)$. The conditions are as follows:

(1) vector $C = [c_0, c_1, ..., c_{N-1}]$ satisfies the property **U** explained above,

(2) $g(x)h(x) = xN - 1$, where $h(x)$ is a monic irreducible polynomial of degree n , guarantee the existence of a *p-ary*, linear cyclic code $\bar{K}$: of blocksize N , such that the augmented code $K = [0, \bar{K}]$ is the Hadamard exponent, for Hadamard matrix $H(p, p_n) = xK$, with $x = e^{2\pi i/p}$, where the core of H is cyclic matrix.

Proof:

First, we note that since $g(x)$ is monic, it divides x^{N-1} , and has degree $= N - n$. Now, we need to show that the matrix E_c whose rows are the nonzero codewords, constitutes a cyclic core for some complex Hadamard matrix H .

Given: we know that C satisfies property U. Hence, all of the nonzero residues of $GF(p)$ lie in C. By cycling through C , we get the desired exponent matrix E_c where we can get every codeword in E_c by cycling the first codeword. (This is because the sequence obtained by cycling through C is an M-invariant sequence.)

We also see that augmentation of each codeword of E_c by adding a leading zero element produces a vector which satisfies Property U. Also, since the code is linear, the mod p vector difference of two arbitrary codewords is also a codeword and thus satisfy Property U. Therefore, the row vectors of the augmented code K form a Hadamard exponent. Thus, xK is the standard form of some complex Hadamard matrix H .

Thus from the above property, we see that the core of E is a circulant matrix consisting of all the $N = p^k - 1$ cyclic shifts of its first row. Such a core is called a cyclic core where in each element of $\mathbb{Z}_p$ appears in each row of E exactly $(N+1)/p = p^{k-1}$ times, and the Hamming distance between any two rows is exactly $(N+1)(p-1)/p = (p-1)p^{k-1}$. The N rows of the core E form a *constant-composition code* - one consisting of N cyclic shifts of some length N over the set $\mathbb{Z}_p$. Hamming distance between any two codewords in $\mathbb{Z}_p$ is $(p-1)p^{k-1}$.

The following can be inferred from the theorem as explained above. (For more detailed reading, the reader is referred to the paper by Heng and Cooke.[4]) *Let $N = p^k - 1$ for p prime and $k \in \mathbb{Z}^+$. Let $g(x) = c_0 + c_1 x + c_2 x^2 + ... + c_{N-k} x^{N-k}$ be a monic polynomial over $\mathbb{Z}_p$, of degree N - k such that $g(x)h(x) = x^N - 1$ over $\mathbb{Z}_p$, for some monic irreducible polynomial $h(x) \in \mathbb{Z}_p[x]$. Suppose that the vector $(c_0, c_1, ..., c_{N-k}, c_{N-k+1}, ..., c_{N-1})$, with $c_i = 0$ for $(N - k) < i < N$, has the property that it contains each element of $\mathbb{Z}_p$ the same number of times. Then, the N cyclic shifts of the vector $g = (c_0, c_1, ..., c_{N-1})$ form the core of the exponent matrix of some Hadamard matrix .*

DNA codes with constant GC-content can obviously be constructed from constant-composition codes (A constant composition code over a k-ary alphabet has the property that the numbers of occurrences of the k symbols within a codeword is the same for each codeword) over $\mathbb{Z}_p$ by mapping the symbols of $\mathbb{Z}_p$ to the symbols of the DNA alphabet, $\mathcal{Q} = \{A, T, C, G\}$. For example, using cyclic constant composition code of length $3^k - 1$ over $\mathbb{Z}_3$ guaranteed by the theorem proved above and the resulting property, and using the mapping that takes 0 to A , 1 to T and 2 to G , we obtain a DNA code $\mathcal{D}$ with $3^k - 1$ and a GC-content of 3^{k-1} . Clearly $d_H = 2.3^{k-1}$ and in fact since $\bar{G} = C$ and no codeword in $\mathcal{D}$ contains no symbol C , we also have $d_H^{RC}(\mathcal{D}) \geq 3^{k-1}$. This is summarized in the following corollary.[4]

Corollary

For any $k \in \mathbb{Z}^+$, there exists DNA codes $\mathbb{D}$ with $3^k - 1$ codewords of length $3^k - 1$, constant GC-content 3^{k-1} , $d_H^{RC}(\mathbb{D}) \geqslant 3^{k-1}$ and in which every codeword is a cyclic shift of a fixed generator codeword g .

Each of the following vectors generates a cyclic core of a Hadamard matrix $H(p, p^n)$ (where $N + 1 = p^n$, and $n = 3$ in this example):[4]

$g^{(1)} = (2220122120200111021121210200)$;

$g^{(2)} = (2021221022200101211201110110)$.

Where, $g(x) = a_0 + a_1 x + ... + a_n x^n$.

Thus, we see how DNA codes can be obtained from such generators by mapping $0, 1, 2$ onto A, T, G . The actual choice of mapping plays a major role in secondary structure formations in the codewords.

We see that all such mappings yield codes with essentially the same parameters. However the actual choice of mapping has a strong influence on the secondary structure of the codewords. For example, the codeword illustrated was obtained from $g^{(1)}$ via the mapping $0 - A; 1 - T; 2 - G$, while the codeword $g^{(2)}$ was obtained from the same generator $g^{(1)}$ via the mapping $0 - G; 1 - T; 2 - A$.

2.3.2   2. Code construction via a Binary Mapping

Perhaps a simpler approach to building/designing DNA codewords is by having a binary mapping by looking at the design problem as that of constructing the codewords as binary codes. i.e. map the DNA codeword alphabet $\mathcal{Q}$ onto the set of 2-bit length binary words as shown: $A \to 00$, $T \to 01$, $C \to 10$, $G \to 11$.

As we can see, the first bit of a binary image clearly determines which complementary pair it belongs to.

Let q be a DNA sequence. The sequence $b(q)$ obtained by applying the mapping given above to q, is called the *binary image* of q.

Now, let $b(q) = b_0 b_1 b_2 ... b_{2n-1}$.

Now, let the subsequence $e(q) = b_0 b_2 ... b_{2n-2}$ be called the even subsequence of $b(q)$, and $o(q) = b_1 b_3 b_5 ... b_{2n-1}$ be called the odd subsequence of $b(q)$.

Thus, for example, for $q = ACGTCC$, then, $b(q) = 001011011010$.

$e(q)$ will then be $= 011011$ and $o(q) = 001100$.

Let us define an *even component* as $\mathcal{E}(\mathcal{C}) = \{e(x) : x \in \mathcal{C}\}$, and an *odd component* as $\mathcal{O}(\mathcal{C}) = \{o(x) : x \in \mathcal{C}\}$.

From this choice of binary mapping, the GC-content of DNA sequence $q =$ Hamming weight of $e(q)$.

Hence, a DNA code $\mathcal{C}$ is a constant GC-content codeword if and only if its even component $\mathcal{E}(\mathcal{C})$ is a constant-weight code.

Let $\mathcal{B}$ be a binary code consisting of M codewords of length n and minimum distance d_{min}, such that $c \in \mathcal{B}$ implies that $\bar{c} \in \mathcal{B}$.

For $w > 0$, consider the constant-weight subcode $\mathcal{B}_w = \{u \in \mathcal{B} : w_H(u) = w\}$, where $w_H(.)$ denotes Hamming weight. Choose $w > 0$ such that $n \geq 2w + \lceil d_{min}/2 \rceil$, and consider a DNA code, $\mathcal{C}_w$, with the following choice for its even and odd components:

$$\mathcal{E} = \{a\bar{b} : a, b \in \mathcal{B}_w\}, \quad \mathcal{O} = \{ab^{RC} : a, b \in \mathcal{B}, a <_{lex} b\}.$$

Where $<_{lex}$ denotes lexicographic ordering. The $a <_{lex} b$ in the definition of $\mathcal{O}$ ensures that if $ab^{RC} \in \mathcal{O}$, then $ba^{RC} \notin \mathcal{O}$, so that distinct codewords in $\mathcal{O}$ cannot be reverse-complements of each other.

The code $\mathcal{E}_w$ has $|\mathcal{B}_w|^2$ codewords of length $2n$ and constant weight n.

Furthermore, $d_H(\mathcal{E}_w \geq d_{min})$ and $d_H^R(\mathcal{E}_w \geq d_{min})$ (this is because $\mathcal{B}_w$ is a subset of the codewords in $\mathcal{B}$).

Also, $d_H(a\bar{b}, d^{RC}c^R) = d_H(a, d^{RC}) + d_H(\bar{b}, c^R) = d_H(a, d^{RC}) + d_H(c, b^{RC})$.

Note that b and d both have weight w. This implies that b^{RC} and d^{RC} have weight $n - w$.

And due to the weight constraint on w, we must have for all $a, b, c, d \in \mathcal{B}_w$, $d_H(a\bar{b}, d^{RC}c^R) \geq 2\lceil d_{min}/2 \rceil \geq d_{min}$.

Thus, the code $\mathcal{O}$ has $M(M-1)/2$ codewords of length $2n$.

From this, we see that $d_H((O)) \geq d_{min}$ (because of the fact that the component codewords of (O) are taken from $\mathcal{B}$). Similarly, $d_H^{RC}((O)) \geq d_{min}$.

Therefore, the DNA code

$$\mathcal{C} = \bigcup_{w=d_{min}}^{w_{max}} \mathcal{C}_w$$

with $w_{max} = (n - \lceil d_{min}/2 \rceil)/2$, has $\frac{1}{2}M(M-1)\sum_{w=d_{min}}^{w_{max}} |A_w^2|$ codewords of length $2n$, and satisfies $d_H(\mathcal{B}) \geq d_{min}$ and $d_H^{RC}(\mathcal{B}) \geq d_{min}$.

From the examples listed above, one can wonder what could be the future potential of DNA-based computers?

Despite its enormous potential, this method is highly unlikely to be implemented in home computers or even computers at offices, etc. because of the sheer flexibility and speed as well as cost factors that favor silicon chip based devices used for the computers today.[2]

However, such a method could be used in situations where the only available method is this and requires the accuracy associated with the DNA hybridization mechanism; applications which require operations to be performed with a high degree of reliability.

Currently, there are several software packages, such as the Vienna package,[7] which can predict secondary structure formations in single stranded DNAs (i.e. oligonucleotides) or RNA sequences.

2.4 See also

- Coding theory
- Bioinformatics
- Biocomputers
- Computational gene

2.5 References

[1] Adleman, L. (1994). "Molecular computation of solutions to combinatorial problem" (PDF). *Science* **266** (5187): 1021–4. doi:10.1126/science.7973651. PMID 7973651.

[2] Mansuripur, M.; Khulbe, P.K.; Kuebler, S.M.; Perry, J.W.; Giridhar, M.S.; Peyghambarian, N. (2003). "Information storage and retrieval using macromolecules as storage media". *Optical Society of America Technical Digest Series.*

[3] Milenkovic, Olgica; Kashyap, Navin (14–18 March 2005). *On the Design of codes for DNA computing.* International Workshop on Coding and Cryptography. Bergen, Norway. doi:10.1007/11779360_9.

[4] Cooke, C. (1999). "Polynomial construction of complex Hadamard matrices with cyclic core". *Applied Mathematics Letters* **12**: 87–93. doi:10.1016/S0893-9659(98)00131-1.

[5] Adámek, Jiří (1991). *Foundations of coding: theory and applications of error-correcting codes, with an introduction to cryptography and information theory.* Chichester: Wiley. doi:10.1002/9781118033265. ISBN 978-0-471-62187-4.

[6] Zierler, N. (1959). "Linear recurring sequences". *J. Soc. Indust. Appl. Math.* **7**: 31–48. doi:10.1137/0107003.

[7] "The Vienna RNA secondary structure package".

2.6 External links

- Atri Rudra's course at The State University of New York, Buffalo

Chapter 3

Robert Dirks

Robert Dirks (May 29, 1978 – February 3, 2015) was an American chemist known for his theoretical and experimental work in DNA nanotechnology. Born in Thailand to a Thai Chinese mother and American father, he moved to Spokane, Washington at a young age. Dirks was the first graduate student in Niles Pierce's research group at the California Institute of Technology, where his dissertation work was on algorithms and computational tools to analyze nucleic acid thermodynamics and predict their structure. He also performed experimental work developing a biochemical chain reaction to self-assemble nucleic acid devices. Dirks later worked at D. E. Shaw Research on algorithms for protein folding that could be used to design new pharmaceuticals.

In February 2015, Dirks died in the Valhalla train crash, the deadliest accident in the history of Metro-North Railroad. An award for early-career achievement in molecular programming research was established in his honor.

3.1 Early life

Dirks was born in Bangkok, Thailand, in 1978.[1] His father, Michael Dirks, was a mathematics teacher at the International School Bangkok recruited from the United States, his mother Suree a Thai Chinese woman who worked in a bank at the time.[2] After about a year, the family, including older brother William, moved back to Vancouver, British Columbia, Canada so that his father could pursue doctoral studies in mathematics education at the University of British Columbia.[3] Four years later the family settled in the elder Dirks' hometown of Spokane, Washington, where he took a job teaching math at North Central High School and Spokane Falls Community College.[1]

Robert attended Lewis and Clark High School, where he excelled academically, entering and winning many math competitions. He was selected to do cardiovascular research at the University of Washington over the summer before his senior year. During that year, he received the top score of 5 on every Advanced Placement exam he took, and was chosen as class valedictorian in 1996.[4] Shortly after graduation Robert and three of his classmates learned that their entry in the ExploraVision national scientific contest was one of three high school winners earning them and their families a trip to Washington, D.C. The topic of their project was the future of nanotechnology.[5]

Although he had been accepted to the Massachusetts Institute of Technology, he chose instead to do his undergraduate work at Wabash College in Crawfordsville, Indiana. There he did a double major in chemistry and math and a double minor in biology and music. In the latter area, he played bassoon, clarinet and piano.[4] After graduating *summa cum laude*,[6] and with Phi Beta Kappa honors, from Wabash in 2000, he was accepted at the California Institute of Technology in Pasadena, California where he began graduate studies in chemistry. He received his doctorate in 2005, and remained at Caltech for a postdoctoral fellowship. During his years there he met Christine Ueda, another doctoral student who became his wife.[1]

3.2 Research

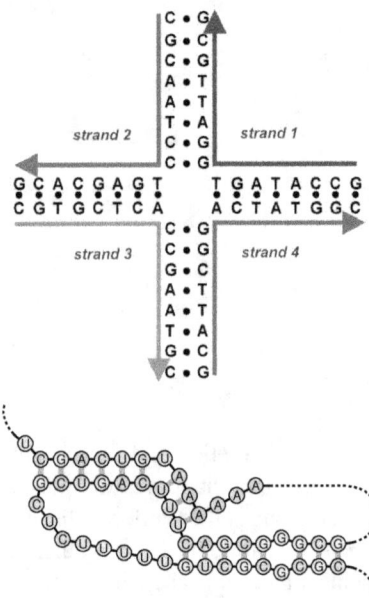

Dirks developed algorithms and computational tools for predicting the structure and thermodynamics of certain classes of nucleic acid structures: those containing multiple strands *(top)* and pseudoknots *(bottom)*.

Dirks was the first graduate student in the laboratory of Niles Pierce at Caltech.[7] His dissertation was entitled "Analysis, design, and construction of nucleic acid devices".[8]

Dirks' work in computational chemistry involved creating algorithms and computational tools for the analysis of nucleic acid thermodynamics and nucleic acid structure prediction.[9] Dirks wrote the initial code for the NUPACK suite of nucleic acid design and analysis tools,[10] which generates base pairing probabilities through calculation of the statistical partition function.[11] Unlike other structure prediction tools, NUPACK is capable of handling an arbitrary number of interacting strands rather than being limited to one or two. Dirks also developed an algorithm capable of efficiently handling certain types of pseudoknots, a class of structure that is more computationally intensive to analyze, although NUPACK only implements this ability for single RNA strands.[11][12][13]

His experimental work pioneered the hybridization chain reaction method, the first demonstration of the self-assembly of nucleic acid structures conditional on a molecular input.[9] The method arose from attempts to use DNA hairpins as "fuel" for DNA machines, but Dirks and Pierce realized that they could instead be used for signal amplification, and when used in conjunction with an aptamer, as a biosensor.[14] As an enzyme-free, isothermal method, it later found application as the basis of an immunoassay method,[15] for *in situ* hybridization imaging of gene expression,[16] and as the basis for catalytic, isothermal self-assembly of DNA nanostructures.[17]

Dirks then worked at D. E. Shaw Research in Manhattan to develop methods for computational protein structure prediction[18] for the design of new drugs, beginning in 2006.[1]

3.3 Later life and death

Dirks and Ueda married in 2007. She initially also worked at D. E. Shaw Research, but stopped in 2010 to raise the first of what would become two children. The couple settled in the Westchester County suburb of Chappaqua, New York, near the train station. He rose early to commute to his job via Metro-North Railroad's Harlem Line, and returned late but devoted as much time as possible on evenings and weekends to his children.[4]

On February 3, 2015, he was riding home in the front car of his train, which his brother says he likely did to take advantage of the quieter atmosphere,[2] when it struck an SUV at a grade crossing north of Valhalla, 5 miles (8.0 km) south of

Chappaqua. The train dragged the SUV while it came to a stop, loosening segments of the third rail that accumulated in the front car. Dirks was killed, along with the SUV driver and four other passengers, making it the deadliest accident in Metro-North's history.[19]

Reactions to his death came from many quarters, many paying tribute to his scientific prowess. His father recalled that "he always got everything the first time. He always excelled." Greg Sampson, Dirks' math teacher at Lewis and Clark, remembered when his student had finished an advanced class in trigonometry in just two weeks, something no other student of his has ever done, saying "he was just an amazing, amazing student." Niles Pierce recalled how Dirks was willing to take a chance on working with a younger professor. His former postdoc was, he said, "an unusual student, even for Caltech... He did remarkable things."[4] D. E. Shaw Research, his employer, called him "a brilliant scientist who made tremendous contributions to our own research, and to the broader scientific community."[1]

In April 2015, the International Society for Nanoscale Science, Computation, and Engineering (ISNCSE), the main scientific society for DNA nanotechnology and DNA computing, established the Robert Dirks Molecular Programming Prize to recognize early-career scientists for molecular programming research.[20]

3.4   Notable works

- Dirks, R. M.; Pierce, N. A. (2003). "A partition function algorithm for nucleic acid secondary structure including pseudoknots". *Journal of Computational Chemistry* **24** (13): 1664. doi:10.1002/jcc.10296.

- Dirks, R. M.; Pierce, N. A. (2004). "Triggered amplification by hybridization chain reaction". *Proceedings of the National Academy of Sciences* **101** (43): 15275. doi:10.1073/pnas.0407024101.

- Dirks, R. M.; Bois, J. S.; Schaeffer, J. M.; Winfree, E.; Pierce, N. A. (2007). "Thermodynamic Analysis of Interacting Nucleic Acid Strands". *SIAM Review* **49**: 65. doi:10.1137/060651100.

- Piana, S.; Lindorff-Larsen, K.; Dirks, R. M.; Salmon, J. K.; Dror, R. O.; Shaw, D. E. (2012). "Evaluating the Effects of Cutoffs and Treatment of Long-range Electrostatics in Protein Folding Simulations". *PLoS ONE* **7** (6): e39918. doi:10.1371/journal.pone.0039918.

3.5   See also

- List of chemists

- List of people from Spokane, Washington

- List of Wabash College people

- List of California Institute of Technology people

3.6   References

[1] Miller, Stephen (February 5, 2015). "Robert Dirks, Scientist at D.E. Shaw Research, Dies at 36". *Bloomberg News*. Retrieved April 5, 2015.

[2] Armaghan, Sarah (February 5, 2015). "Robert Michael Dirks, one of five passengers killed in Metro-North crash, was accomplished scientist and devoted father". *Newsday*. Retrieved April 6, 2015.

[3] "A life begun in Bangkok ends in fiery New York train wreck". *Bangkok Post*. February 6, 2015. Retrieved April 5, 2015.

[4] Culver, Nina (February 5, 2015). "Robert Dirks, the LC grad who died in train crash, was a family man with 'unusual talent'". *The Spokesman-Review* (Spokane, WA). Retrieved April 5, 2015.

[5] Fernandez, Maisy. "Francoise Kuester Chosen Gu's Teacher Of The Year". The Spokesman-Review. Retrieved 2015-07-13.

[6] "Robert Dirks". D. E. Shaw Research. Retrieved April 6, 2015.

[7] Culver, Nina (5 February 2015). "Robert Dirks, the LC grad who died in train crash, was a family man with 'unusual talent'". *The Spokesman-Review*. Retrieved 27 February 2015.

[8] Dirks, Robert (2005). "Analysis, design, and construction of nucleic acid devices". California Institute of Technology. Retrieved 27 February 2015.

[9] "The Pierce Lab". California Institute of Technology. Archived from the original on 11 February 2015. Retrieved 26 February 2015.

[10] "NUPACK: Nucleic Acid Package". California Institute of Technology. Archived from the original on 23 February 2015. Retrieved 26 February 2015.

[11] Andersen, Ebbe Sloth (2010). "Prediction and design of DNA and RNA structures". *New Biotechnology* **27** (3): 184–193. doi:10.1016/j.nbt.2010.02.012.

[12] Bindewald, Eckart; Afonin, Kirill; Jaeger, Luc; Shapiro, Bruce A. (2011). "Multi-Strand RNA Secondary Structure Prediction and Nanostructure Design including Pseudoknots". *ACS Nano* **5** (12): 9542–9551. doi:10.1021/nn202666w. PMC 3263976.

[13] Schroeder, Susan J. (2009). "Advances in RNA Structure Prediction from Sequence: New Tools for Generating Hypotheses about Viral RNA Structure-Function Relationships". *Journal of Virology* **83** (13): 6326–6334. doi:10.1128/JVI.00251-09.

[14] Evenko, Daniel (2004). "Hybridization chain reaction". *Nature Methods* **1**: 186–187. doi:10.1038/nmeth1204-186a.

[15] Deng, Y.; Nie, J.; Zhang, X. H.; Zhao, M. Z.; Zhou, Y. L.; Zhang, X. X. (2014). "Hybridization chain reaction-based fluorescence immunoassay using DNA intercalating dye for signal readout". *The Analyst* **139** (13): 3378. doi:10.1039/C4AN00190G.

[16] Choi, H. M. T.; Beck, V. A.; Pierce, N. A. (2014). "Next-Generation *in Situ* Hybridization Chain Reaction: Higher Gain, Lower Cost, Greater Durability". *ACS Nano*: 140408151851004. doi:10.1021/nn405717p.

[17] Yin, P.; Choi, H. M. T.; Calvert, C. R.; Pierce, N. A. (2008). "Programming biomolecular self-assembly pathways". *Nature* **451** (7176): 318. doi:10.1038/nature06451.

[18] Yee, Vivian (5 February 2015). "The Lives of 3 Crash Victims Who Shared a Metro-North Routine". *The New York Times*. Retrieved 27 February 2015.

[19] Santora, Marc; Flegenheimer, Matt (February 4, 2015). "Investigation Underway in Metro-North Train Crash". *New York Times*. Retrieved April 7, 2015.

[20] "Robert Dirks '00 memorial prize announced". Wabash College. 1 May 2015. Retrieved 5 May 2015.

Chapter 4

DNA computing

For hypothetical computers using brain-to-brain interfaces, see Biological computer.

DNA computing is a branch of computing which uses DNA, biochemistry, and molecular biology hardware, instead of the traditional silicon-based computer technologies. DNA computing—or, more generally, biomolecular computing—is a fast-developing interdisciplinary area. Research and development in this area concerns theory, experiments, and applications of DNA computing. The term "molectronics" has sometimes been used, but this term had already been used for an earlier technology, a then-unsuccessful rival of the first integrated circuits;[1] this term has also been used more generally, for molecular-scale technology.[2]

4.1 History

This field was initially developed by Leonard Adleman of the University of Southern California, in 1994.[3] Adleman demonstrated a proof-of-concept use of DNA as a form of computation which solved the seven-point Hamiltonian path problem. Since the initial Adleman experiments, advances have been made and various Turing machines have been proven to be constructible.[4][5]

While the initial interest was in using this novel approach to tackle NP-hard problems, it was soon realized that they may not be best suited for this type of computation, and several proposals have been made to find a "killer application" for this approach. In 1997, computer scientist Mitsunori Ogihara working with biologist Animesh Ray suggested one to be the evaluation of Boolean circuits and described an implementation.[6][7]

In 2002, researchers from the Weizmann Institute of Science in Rehovot, Israel, unveiled a programmable molecular computing machine composed of enzymes and DNA molecules instead of silicon microchips.[8] On April 28, 2004, Ehud Shapiro, Yaakov Benenson, Binyamin Gil, Uri Ben-Dor, and Rivka Adar at the Weizmann Institute announced in the journal Nature that they had constructed a DNA computer coupled with an input and output module which would theoretically be capable of diagnosing cancerous activity within a cell, and releasing an anti-cancer drug upon diagnosis.[9]

In January 2013, researchers were able to store a JPEG photograph, a set of Shakespearean sonnets, and an audio file of Martin Luther King, Jr.'s speech I Have a Dream on DNA digital data storage.[10]

In March 2013, researchers created a transcriptor (a biological transistor).

4.2 Idea

The organisation and complexity of all living beings is based on a coding system functioning with four key components of the DNA-molecule. Because of this, the DNA is very suited as a medium for data processing.[11] According to different calculations a DNA-computer with one liter of fluid containing six grams of DNA could potentially have a memory

30

capacity of 3072 exabytes. On top of that, the theoretical maximum data transfer speed would be enormous due to the massive parallelism of the calculations. Therefore, about 1000 petaFLOPS could be reached, while today's most powerful computers do not go above a few dozen (33.86 petaFLOPS by Tianhe-2 being the current record holder).

4.3 Pros and cons

The slow processing speed of a DNA-computer (the response time is measured in minutes, hours or days, rather than milliseconds) is compensated by its potential to make a high amount of multiple parallel computations. This allows the system to take a similar amount of time for a complex calculation as for a simple one. This is achieved by the fact that millions or billions of molecules interact with each other simultaneously. However, it is a lot harder to analyse the answers given by a DNA-Computer than by a digital one.

4.4 Examples/Prototypes

In 1994 Leonard Adleman presented the first prototype of a DNA-Computer. The TT-100 was a test tube filled with 100 microliters of a DNA-solution. He managed to solve for example an instance of the directed Hamiltonian path problem.[12]

In another experiment a simple version of the "travelling salesman problem" was "solved". For this purpose, different DNA-fragments were created, each one of them representing a city that had to be visited. Every one of these fragments is capable of a linkage with the other fragments created. These DNA-fragments were produced and mixed in a test tube. Within seconds, the small fragments form bigger ones, representing the different travel routes. Through a chemical reaction (that lasts a few days), the DNA-fragments representing the longer routes were eliminated. The remains are the solution to the problem. However, because of technical restraints of our day and age, it is impossible to evaluate the results. Therefore, the experiment isn't suitable for application, but it is nevertheless a proof of concept.

4.4.1 Combinatorial problems

First results to these problems were obtained by Leonard Adleman (NASA JPL)

- In 1994: Solving a Hamiltonian path in a graph with 7 summits.

- In 2002: Solving a NP-complete problem as well as a 3-SAT problem with 20 variables.

4.4.2 Tic Tac Toe game

In 2002, J. Macdonald, D. Stefanovic and Mr. Stojanovic created a DNA computer able to play Tic-tac-toe against a human player.[13] The calculator consists of nine bins corresponding to the nine squares of the game. Each bin contains a substrate and various combinations of DNA enzymes. The substrate itself is composed of a DNA strand onto which was grafted a fluorescent chemical group at one end, and the other end, a repressor group. Fluorescence is only active if the molecules of the substrate are halved. The DNA enzyme simulate logical functions. For example, such a DNA will unfold if we introduce two specific types of DNA strand, reproducing the logic function AND.

By default, the computer is supposed to play first in the central square. The human player has then as a starter eight different types of DNA strands assigned to each of eight boxes that may be played. To indicate that box nr. i is being ticked, the human player pours into all bins the strands corresponding to input #i. These strands bind to certain DNA enzymes present in the bins, resulting in one of these two bins in the deformation of the DNA enzymes which binds to the substrate and cuts it. The corresponding bin becomes fluorescent, indicating which box is being played by the DNA computer. The various DNA enzymes are divided into various bins in such a way to ensure the victory of the DNA computer against the human player.

4.5   Capabilities

DNA computing is a form of parallel computing in that it takes advantage of the many different molecules of DNA to try many different possibilities at once.[14] For certain specialized problems, DNA computers are faster and smaller than any other computer built so far. Furthermore, particular mathematical computations have been demonstrated to work on a DNA computer. As examples, DNA molecules have been utilized to tackle the assignment problem[15] and GPS mapping.[16] Aran Nayebi[17] has provided a general implementation of Strassen's matrix multiplication algorithm on a DNA computer, although there are problems with scaling. In addition, Caltech researchers have created a circuit made from 130 unique DNA strands, which is able to calculate the square root of numbers up to 15.[18]

DNA computing does not provide any new capabilities from the standpoint of computability theory, the study of which problems are computationally solvable using different models of computation. For example, if the space required for the solution of a problem grows exponentially with the size of the problem (EXPSPACE problems) on von Neumann machines, it still grows exponentially with the size of the problem on DNA machines. For very large EXPSPACE problems, the amount of DNA required is too large to be practical.

4.6   Methods

There are multiple methods for building a computing device based on DNA, each with its own advantages and disadvantages. Most of these build the basic logic gates (AND, OR, NOT) associated with digital logic from a DNA basis. Some of the different bases include DNAzymes, deoxyoligonucleotides, enzymes, toehold exchange.

4.6.1   DNAzymes

Catalytic DNA (deoxyribozyme or DNAzyme) catalyze a reaction when interacting with the appropriate input, such as a matching oligonucleotide. These DNAzymes are used to build logic gates analogous to digital logic in silicon; however, DNAzymes are limited to 1-, 2-, and 3-input gates with no current implementation for evaluating statements in series.

The DNAzyme logic gate changes its structure when it binds to a matching oligonucleotide and the fluorogenic substrate it is bonded to is cleaved free. While other materials can be used, most models use a fluorescence-based substrate because it is very easy to detect, even at the single molecule limit.[19] The amount of fluorescence can then be measured to tell whether or not a reaction took place. The DNAzyme that changes is then "used," and cannot initiate any more reactions. Because of this, these reactions take place in a device such as a continuous stirred-tank reactor, where old product is removed and new molecules added.

Two commonly used DNAzymes are named E6 and 8-17. These are popular because they allow cleaving of a substrate in any arbitrary location.[20] Stojanovic and MacDonald have used the E6 DNAzymes to build the MAYA I[21] and MAYA II[22] machines, respectively; Stojanovic has also demonstrated logic gates using the 8-17 DNAzyme.[23] While these DNAzymes have been demonstrated to be useful for constructing logic gates, they are limited by the need for a metal cofactor to function, such as Zn^{2+} or Mn^{2+}, and thus are not useful in vivo.[19][24]

A design called a *stem loop*, consisting of a single strand of DNA which has a loop at an end, are a dynamic structure that opens and closes when a piece of DNA bonds to the loop part. This effect has been exploited to create several logic gates. These logic gates have been used to create the computers MAYA I and MAYA II which can play tic-tac-toe to some extent.[25]

4.6.2   Enzymes

Enzyme based DNA computers are usually of the form of a simple Turing machine; there is analogous hardware, in the form of an enzyme, and software, in the form of DNA.[26]

Benenson, Shapiro and colleagues have demonstrated a DNA computer using the FokI enzyme[27] and expanded on their work by going on to show automata that diagnose and react to prostate cancer: under expression of the genes PPAP2B and GSTP1 and an over expression of PIM1 and HPN.[9] Their automata evaluated the expression of each gene, one gene at a time, and on positive diagnosis then released a single strand DNA molecule (ssDNA) that is an antisense for MDM2. MDM2 is a repressor of protein 53, which itself is a tumor suppressor.[28] On negative diagnosis it was decided to release a suppressor of the positive diagnosis drug instead of doing nothing. A limitation of this implementation is that two separate automata are required, one to administer each drug. The entire process of evaluation until drug release took around an hour to complete. This method also requires transition molecules as well as the FokI enzyme to be present. The requirement for the FokI enzyme limits application *in vivo*, at least for use in "cells of higher organisms".[29] It should also be pointed out that the 'software' molecules can be reused in this case.

4.6.3 Toehold exchange

DNA computers have also been constructed using the concept of toehold exchange. In this system, an input DNA strand binds to a sticky end, or toehold, on another DNA molecule, which allows it to displace another strand segment from the molecule. This allows the creation of modular logic components such as AND, OR, and NOT gates and signal amplifiers, which can be linked into arbitrarily large computers. This class of DNA computers does not require enzymes or any chemical capability of the DNA.[30]

4.6.4 Algorithmic self-assembly

Main article: DNA nanotechnology: Algorithmic self-assembly

DNA nanotechnology has been applied to the related field of DNA computing. DNA tiles can be designed to contain multiple sticky ends with sequences chosen so that they act as Wang tiles. A DX array has been demonstrated whose assembly encodes an XOR operation; this allows the DNA array to implement a cellular automaton which generates a fractal called the Sierpinski gasket. This shows that computation can be incorporated into the assembly of DNA arrays, increasing its scope beyond simple periodic arrays.[31]

4.7 Alternative technologies

A partnership between IBM and CalTech was established in 2009 aiming at "DNA chips" production.[32] A CalTech group is working on the manufacturing of these nucleic-acid-based integrated circuits. One of these chips can compute whole square roots.[33] A compiler has been written[34] in Perl.

4.8 See also

- Biocomputers

- Computational gene

- DNA code construction

- DNA sequencing

- Molecular electronics

- Peptide computing

- Parallel computing

- Quantum computing

- Transcriptor

- Wetware computer

- Bioinformatics

- Carbon Nano Tube

4.9   References

[1]   "Molectronic Computer Shown by Texas Instr.", unknown publication, circa 1963, in Box 2, Folder 3, listed in *Jack Kilby Papers: A Guide to the Collection*, Southern Methodist University. .

[2]   "Application-specific methods for testing molectronic or nanoscale devices" (filed April 1, 2004), Patent US 7219314 B1. .

[3]   Adleman, L. M. (1994). "Molecular computation of solutions to combinatorial problems". *Science* **266** (5187): 1021–1024. Bibcode:1994Sci...266.1021A. doi:10.1126/science.7973651. PMID 7973651. — The first DNA computing paper. Describes a solution for the directed Hamiltonian path problem. Also available here:

[4]   Boneh, D.; Dunworth, C.; Lipton, R. J.; Sgall, J. Í. (1996). "On the computational power of DNA". *Discrete Applied Mathematics* **71**: 79–94. doi:10.1016/S0166-218X(96)00058-3. — Describes a solution for the boolean satisfiability problem. Also available here:

[5]   Lila Kari, Greg Gloor, Sheng Yu (January 2000). "Using DNA to solve the Bounded Post Correspondence Problem". *Theoretical Computer Science* **231** (2): 192–203. doi:10.1016/s0304-3975(99)00100-0. — Describes a solution for the bounded Post correspondence problem, a hard-on-average NP-complete problem. Also available here:

[6]   M. Ogihara and A. Ray, "Simulating Boolean circuits on a DNA computer". Algorithmica 25:239–250, 1999.

[7]   "In Just a Few Drops, A Breakthrough in Computing", *New York Times*, May 21, 1997

[8]   Lovgren, Stefan (2003-02-24). "Computer Made from DNA and Enzymes". *National Geographic*. Retrieved 2009-11-26.

[9]   Benenson, Y.; Gil, B.; Ben-Dor, U.; Adar, R.; Shapiro, E. (2004). "An autonomous molecular computer for logical control of gene expression". *Nature* **429** (6990): 423–429. Bibcode:2004Natur.429..423B. doi:10.1038/nature02551. PMC 3838955. PMID 15116117.. Also available here: An autonomous molecular computer for logical control of gene expression

[10] DNA stores poems, a photo and a speech | Science News

[11] Amos, Martyn, et al. "Topics in the theory of DNA computing." *Theoretical computer science* 287.1 (2002): 3-38.

[12] Braich, Ravinderjit S., et al. "Solution of a satisfiability problem on a gel-based DNA computer." *DNA Computing*. Springer Berlin Heidelberg, 2001. 27-42.

[13] [FR] - J. Macdonald, D. Stefanovic et M. Stojanovic, *Des assemblages d'ADN rompus au jeu et au travail*, Pour la Science, Template:N°, January 2009, p. 68-75

[14] Lewin, D. I. (2002). "DNA computing". *Computing in Science & Engineering* **4** (3): 5–8. doi:10.1109/5992.998634.

[15] Shu, Jian-Jun; Wang, Q.-W.; Yong, K.-Y. (2011). "DNA-based computing of strategic assignment problems". *Physical Review Letters* **106** (18): 188702. Bibcode:2011PhRvL.106r8702S. doi:10.1103/PhysRevLett.106.188702.

[16] Shu, Jian-Jun; Wang, Q.-W.; Yong, K.-Y.; Shao F.,; Lee K.J. (2015). "Programmable DNA-mediated multitasking processor". *Journal of Physical Chemistry B* **119** (17): 5639–5644. doi:10.1021/acs.jpcb.5b02165.

[17] Nayebi, Aran (2009). "Fast matrix multiplication techniques based on the Adleman-Lipton model". *arXiv: 0912.0750*: 1–13. External link in |work= (help)

[18] Science NewsFlexbile DNA computer finds square roots

[19] Weiss, S. (1999). "Fluorescence Spectroscopy of Single Biomolecules". *Science* **283** (5408): 1676–1683. Bibcode:1999Sci...283.1676W. doi:10.1126/science.283.5408.1676. PMID 10073925.. Also available here: http://www.lps.ens.fr/~{}vincent/smb/PDF/weiss-1.pdf

[20] Santoro, S. W.; Joyce, G. F. (1997). "A general purpose RNA-cleaving DNA enzyme". *Proceedings of the National Academy of Sciences* **94** (9): 4262–4266. Bibcode:1997PNAS...94.4262S. doi:10.1073/pnas.94.9.4262.. Also available here:

[21] Stojanovic, M. N.; Stefanovic, D. (2003). "A deoxyribozyme-based molecular automaton". *Nature Biotechnology* **21** (9): 1069–1074. doi:10.1038/nbt862. PMID 12923549.. Also available here:

[22] MacDonald, J.; Li, Y.; Sutovic, M.; Lederman, H.; Pendri, K.; Lu, W.; Andrews, B. L.; Stefanovic, D.; Stojanovic, M. N. (2006). "Medium Scale Integration of Molecular Logic Gates in an Automaton". *Nano Letters* **6** (11): 2598–2603. Bibcode:2006NanoL...6.2598M. doi:10.1021/nl0620684. PMID 17090098.. Also available here:

[23] Stojanovic, M. N.; Mitchell, T. E.; Stefanovic, D. (2002). "Deoxyribozyme-Based Logic Gates". *Journal of the American Chemical Society* **124** (14): 3555–3561. doi:10.1021/ja016756v. PMID 11929243.. Also available at

[24] Cruz, R. P. G.; Withers, J. B.; Li, Y. (2004). "Dinucleotide Junction Cleavage Versatility of 8-17 Deoxyribozyme". *Chemistry & Biology* **11**: 57–67. doi:10.1016/j.chembiol.2003.12.012.

[25] Darko Stefanovic's Group, Molecular Logic Gates and MAYA II, a second-generation tic-tac-toe playing automaton.

[26] Shapiro, Ehud (1999-12-07). "A Mechanical Turing Machine: Blueprint for a Biomolecular Computer". Weizmann Institute of Science. Archived from the original on |archive-url= requires |archive-date= (help). Retrieved 2009-08-13.

[27] Benenson, Y.; Paz-Elizur, T.; Adar, R.; Keinan, E.; Livneh, Z.; Shapiro, E. (2001). "Programmable and autonomous computing machine made of biomolecules". *Nature* **414** (6862): 430–434. doi:10.1038/35106533. PMID 11719800.. Also available here:

[28] Bond, G. L.; Hu, W.; Levine, A. J. (2005). "MDM2 is a Central Node in the p53 Pathway: 12 Years and Counting". *Current Cancer Drug Targets* **5** (1): 3–8. doi:10.2174/1568009053332627. PMID 15720184.

[29] Kahan, M.; Gil, B.; Adar, R.; Shapiro, E. (2008). "Towards molecular computers that operate in a biological environment". *Physica D: Nonlinear Phenomena* **237** (9): 1165–1172. Bibcode:2008PhyD..237.1165K. doi:10.1016/j.physd.2008.01.027.. Also available here:

[30] Seelig, G.; Soloveichik, D.; Zhang, D. Y.; Winfree, E. (8 December 2006). "Enzyme-free nucleic acid logic circuits". *Science* **314** (5805): 1585–1588. Bibcode:2006Sci...314.1585S. doi:10.1126/science.1132493. PMID 17158324.

[31] Rothemund, P. W. K.; Papadakis, N.; Winfree, E. (2004). "Algorithmic Self-Assembly of DNA Sierpinski Triangles". *PLoS Biology* **2** (12): e424. doi:10.1371/journal.pbio.0020424. PMC 534809. PMID 15583715.

[32] (CalTech's own article)

[33] Scaling Up Digital Circuit Computation with DNA Strand Displacement Cascades

[34] Online

4.10 Further reading

- Martyn Amos (June 2005). *Theoretical and Experimental DNA Computation*. Springer. ISBN 3-540-65773-8. — The first general text to cover the whole field.

- Gheorge Paun, Grzegorz Rozenberg, Arto Salomaa (October 1998). *DNA Computing - New Computing Paradigms*. Springer-Verlag. ISBN 3-540-64196-3. — The book starts with an introduction to DNA-related matters, the basics of biochemistry and language and computation theory, and progresses to the advanced mathematical theory of DNA computing.

- JB. Waldner (January 2007). *Nanocomputers and Swarm Intelligence*. ISTE. p. 189. ISBN 2-7462-1516-0.

- Zoja Ignatova, Israel Martinez-Perez, Karl-Heinz Zimmermann (January 2008). *DNA Computing Models*. Springer. p. 288. ISBN 978-0-387-73635-8. — A new general text to cover the whole field.

4.11　External links

- DNA modeled computing

- How Stuff Works explanation

- 'DNA computer' cracks code, Physics Web

- Ars Technica

- NY Times DNA Computer for detecting Cancer

- Bringing DNA computers to life, in Scientific American

- Japanese Researchers store information in bacteria DNA

- International Meeting on DNA Computing and Molecular Programming

- LiveScience.com-How DNA Could Power Computers

Leonard Adleman

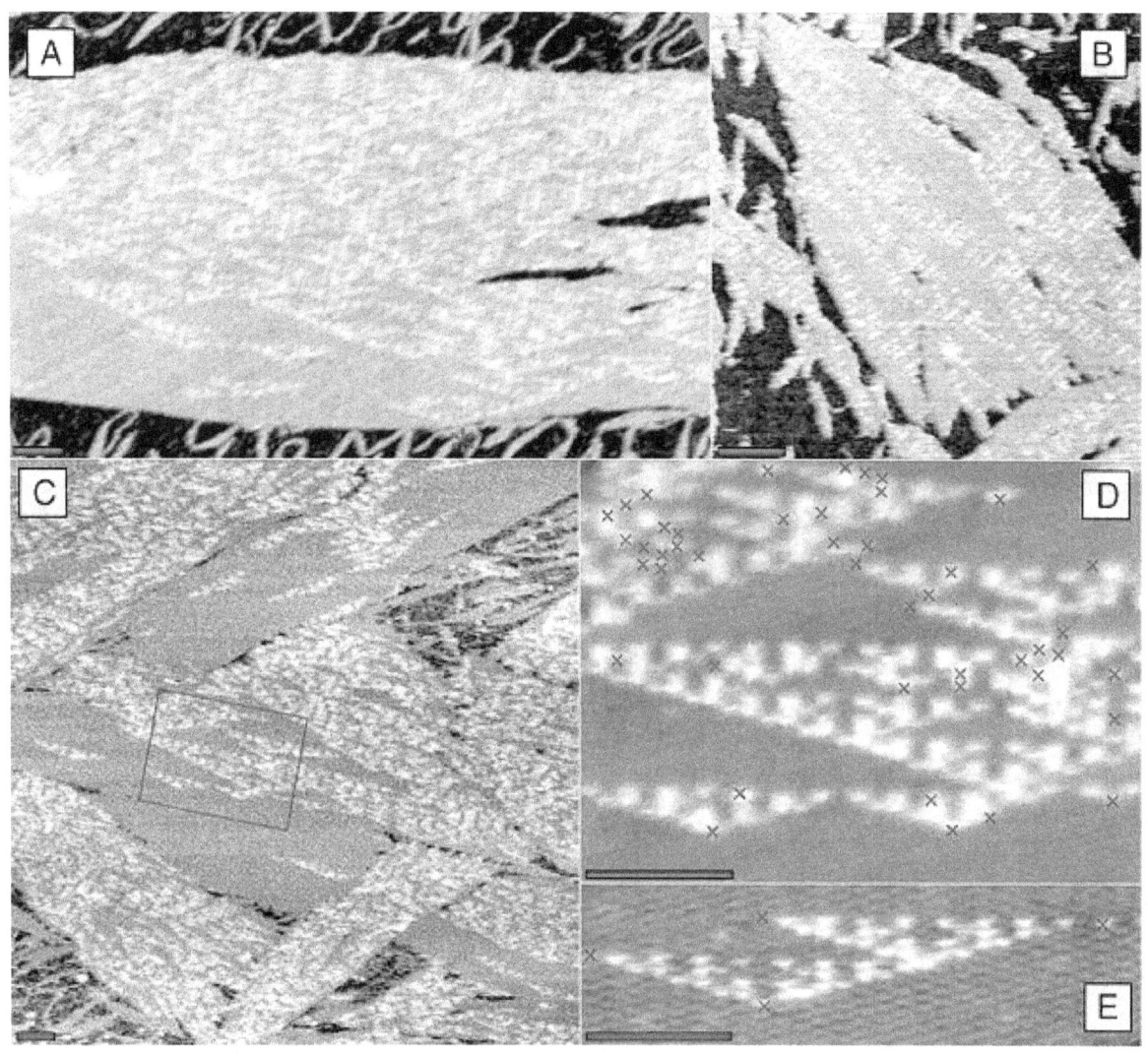

DNA arrays that display a representation of the Sierpinski gasket on their surfaces. Click the image for further details. Image from Rothemund et al., 2004.[31]

Chapter 5

DNA machine

A **DNA machine** is a molecular machine constructed from DNA. Research into DNA machines was pioneered in the late 1980s by Nadrian Seeman and co-workers from New York University. DNA is used because of the numerous biological tools already found in nature that can affect DNA, and the immense knowledge of how DNA works previously researched by biochemists.

DNA machines can be logically designed since DNA assembly of the double helix is based on strict rules of base pairing that allow portions of the strand to be predictably connected based on their sequence. This 'selective stickiness' is a key advantage in the construction of DNA machines.

An example of a DNA machine was reported by Bernard Yurke and co-workers at Lucent Technologies in the year 2000, who constructed molecular tweezers out of DNA.[1] The DNA tweezers contain three strands: A, B and C. Strand A latches onto half of strand B and half of strand C, and so it joins them all together. Strand A acts as a hinge so that the two 'arms' — AB and AC — can move. The structure floats with its arms open wide. They can be pulled shut by adding a fourth strand of DNA (D) 'programmed' to stick to both of the dangling, unpaired sections of strands B and C. The closing of the tweezers was proven by tagging strand A at either end with light-emitting molecules that do not emit light when they are close together. To re-open the tweezers add a further strand (E) with the right sequence to pair up with strand D. Once paired up, they have no connection to the machine BAC, so float away. The DNA machine can be opened and closed repeatedly by cycling between strands D and E. These tweezers can be used for removing drugs from inside fullerenes as well as from a self assembled DNA tetrahedron. The state of the device can be determined by measuring the separation between donor and acceptor fluorophores using FRET.

5.1   References

[1] Yurke B, Turberfield AJ, Mills AP, Simmel FC, Neumann JL (August 2000). "A DNA-fuelled molecular machine made of DNA". *Nature* **406** (6796): 605–8. doi:10.1038/35020524. PMID 10949296.

5.2   See also

- DNA nanotechnology

Chapter 6

DNA origami

DNA origami is the nanoscale folding of DNA to create non-arbitrary two- and three-dimensional shapes at the nanoscale. The specificity of the interactions between complementary base pairs make DNA a useful construction material, through design of its base sequences.[1] DNA is a well-understood material that is suitable for creating scaffolds that hold other molecules in place or to create structures all on its own.

DNA origami was the cover story of *Nature* on March 16, 2006.[2] Since then, DNA origami has progressed past an art form and has found a number of applications from drug delivery systems to uses as circuitry in plasmonic devices; however, most applications remain in a concept or testing phase.[3]

6.1 Overview

The idea of using DNA as a construction material was first introduced in the early 1980s by Nadrian Seeman.[4] The current method of DNA origami was developed by Paul Rothemund at the California Institute of Technology, the process involves the folding of a long single strand of viral DNA aided by multiple smaller "staple" strands.[5] These shorter strands bind the longer in various places, resulting in various shapes, including a smiley face and a coarse map of China and the Americas, along with many three-dimensional structures such as cubes.[6]

To produce a desired shape, images are drawn with a raster fill of a single long DNA molecule. This design is then fed into a computer program that calculates the placement of individual staple strands. Each staple binds to a specific region of the DNA template, and thus due to Watson-Crick base pairing, the necessary sequences of all staple strands are known and displayed. The DNA is mixed, then heated and cooled. As the DNA cools, the various staples pull the long strand into the desired shape. Designs are directly observable via several methods, including Electron Microscopy, atomic force microscopy, or fluorescence microscopy when DNA is coupled to fluorescent materials.[5]

Bottom-up self-assembly methods are considered promising alternatives that offer cheap, parallel synthesis of nanostructures under relatively mild conditions.

Since the creation of this method, software was developed to assist the process using CAD software. This allows researchers to use a computer to determine the way to create the correct staples needed to form a certain shape. One such software called caDNAno is an open source software for creating such structures from DNA. The use of software has not only increased the ease of the process but has also drastically reduced the errors made by manual calculations.[4]

6.2 Applications

Many potential applications have been suggested in literature, including enzyme immobilization, drug carry capsules, and nanotechnological self-assembly of materials. Though DNA is not the natural choice for building active structures for nanorobotic applications, due to its lack of structural and catalytic versatility, several papers have examined the possibility of molecular walkers on origami and switches for algorithmic computing.[6][7] The followings list some of the reported applications conducted in the laboratories with clinical potential.

- Researchers at the Harvard University Wyss Institute reported the self-assembling and self-destructing drug delivery vessels using the DNA origami in the lab tests. The DNA nanorobot they created is an open DNA tube with a hinge on one side which can be clasped shut. The drug filled DNA tube is held shut by DNA aptamer, configured to identify and seek certain diseased related protein. Once the origami nanobots get to the infected cells, the aptamers break apart and release the drug. The first disease model the researchers used was leukemia and lymphoma.[8]

- Researchers in the National Center for Nanoscience and Technology in Beijing and Arizona State University reported a DNA origami delivery vehicle for Doxorubicin, a well-known anti-cancer drug. The drug was non-covalently attached to DNA origami nanostructures through intercalation and a high drug load was achieved. The DNA-Doxorubicin complex was taken up by human breast adenocarcinoma cancer cells (MCF-7) via cellular internalization with much higher efficiency than doxorubicin in free form. The enhancement of cell killing activity was observed not only in regular MCF-7, more importantly, also in doxorubicin-resistant cells. The scientists theorized that the doxorubicin-loaded DNA origami inhibits lysosomal acidification, resulting in cellular redistribution of the drug to action sites, thus increasing the cytotoxicity against the tumor cells.[9][10]

- In a study conducted by a group of scientists from iNANO center and CDNA Center in Aarhus university (Aarhus), researchers were able to construct a small multi-switchable 3D DNA Box Origami. The proposed nanoparticle was characterized by AFM, TEM and FRET. The constructed box was shown to have a unique reclosing mechanism, which enabled it to repeatedly open and close in response to a unique set of DNA or RNA keys. The authors proposed that this "DNA device can potentially be used for a broad range of applications such as controlling the function of single molecules, controlled drug delivery, and molecular computing.".[11]

- Nanorobots made of DNA origami demonstrated computing capacities and completed pre-programmed task inside the living organism was reported by a team of bioengineers at Wyss Institute at Harvard University and Institute of Nanotechnology and Advanced Materials at Bar-Ilan University. As a proof of concept, the team injected various kinds of nanobots (the curled DNA encasing molecules with fluorescent markers) into live cockroaches. By tracking the markers inside the cockroaches, the team found the accuracy of delivery of the molecules (released by the uncurled DNA) in target cells, the interactions among the nanobots and the control are equivalent to a computer system. The complexity of the logic operations, the decisions and actions, increases with the increased number of nanobots. The team estimated that the computing power in the cockroach can be scaled up to that of an 8-bit computer.[12][13]

- DNA is folded into an octahedron and coated with a single bilayer of phospholipid, mimicking the envelope of a virus particle. The DNA nanoparticles, each at about the size of a virion, are able to remain in circulation for hours after injected into mice. It also elicits much lower immune response than the uncoated particles. It presents a potential use in drug delivery, reported by researchers in Wyss Institute at Harvard University.[14][15]

6.3 Similar approaches

The idea of using protein design to accomplish the same goals as DNA origami has surfaced as well. Researchers at the National Institute of Chemistry in Slovenia are working on using rational design of protein folding to create structures much like those seen with DNA origami. The main focus of current research in protein folding design is in the drug delivery field, using antibodies attached to proteins as a way to create a targeted vehicle.[16][17]

6.4 See also

- DNA nanotechnology

- Molecular self-assembly

- Folding@home

6.5   References

[1] **Structural DNA nanotechnology: from design to applications** Zadegan, R.M.; Norton, M.L (2012). "Structural DNA Nanotechnology: From Design to Applications". *Int. J. Mol. Sci* **13**: 7149–7162. doi:10.3390/ijms13067149. PMC 3397516. PMID 22837684.

[2] *Nature, Volume 440 (7082) March 16, 2006*

[3] {http://www.nature.com/news/2010/100310/full/464158a.html 'Nature, Volume 464 March 10, 2010"}

[4] "Rapid prototyping of 3D DNA-origami shapes with caDNAno", "Oxford Journal", May 11, 2009

[5] Rothemund, Paul W. K. (2006). "Folding DNA to create nanoscale shapes and patterns". *Nature* **440** (7082): 297–302. doi:10.1038/nature04586. ISSN 0028-0836. PMID 16541064.

[6] Lin, Chenxiang; Liu, Yan; Rinker, Sherri; Yan, Hao (2006). "DNA Tile Based Self-Assembly: Building Complex Nanoarchitectures". *ChemPhysChem* **7** (8): 1641–7. doi:10.1002/cphc.200600260. PMID 16832805.

[7] *DNA 'organises itself' on silicon*,*BBC News*, August 17, 2009

[8] Garde, Damian (May 15, 2012). "DNA origami could allow for 'autonomous' delivery". fiercedrugdelivery.com. Retrieved May 25, 2012.

[9] "Folded DNA becomes Trojan horse to attack cancer". NewScientist. 18 August 2012. Retrieved 22 August 2012.

[10] Jiang, Qiao; Song, Chen; Nangreave, Jeanette; Liu, Xiaowei; Lin, Lin; Qiu, Dengli; Wang, Zhen-Gang; Zou, Guozhang; Liang, Xingjie; Yan, Hao; Ding, Baoquan (2012). "DNA Origami as a Carrier for Circumvention of Drug Resistance". *Journal of the American Chemical Society* **134** (32): 13396–13403. doi:10.1021/ja304263n.

[11] M. Zadegan, Reza; et. al. (2012). "Construction of a 4 Zeptoliters Switchable 3D DNA Box Origami". *ACS Nano* **6** (11): 10050–10053. doi:10.1021/nn303767b.

[12] Spickernell, Sarah (8 April 2014). "DNA nanobots deliver drugs in living cockroaches". New Scientist. Retrieved 9 June 2014.

[13] Amir, Y; Ben-Ishay, E; Levner, D; Ittah, S; Abu-Horowitz, A; Bachelet, I (2014). "Universal computing by DNA origami robots in a living animal". *Nature Nanotechnology* (Nature) **9** (5): 353–357. doi:10.1038/nnano.2014.58.

[14] Gibney, Michael (23 April 2014). "DNA nanocages that act like viruses bypass the immune system to deliver drugs". fierce-drugdelivery.com. Retrieved 19 June 2014.

[15] Perrault, S; Shih, W (2014). "Virus-Inspired Membrane Encapsulation of DNA Nanostructures To Achieve *In Vivo* Stability". *ACS Nano* (ACS) **8** (5): 5132–5140. doi:10.1021/nn5011914.

[16] Peplow, Mark (28 April 2013). "Protein gets in on DNA's origami act". *Nature*. doi:10.1038/nature.2013.12882.

[17] Zadegan, Reza M.; Norton, Michael L. (June 2012). "Structural DNA Nanotechnology: From Design to Applications". *Int. J. Mol. Sci.* **13** (6): 7149–7162. doi:10.3390/ijms13067149. PMC 3397516. PMID 22837684.

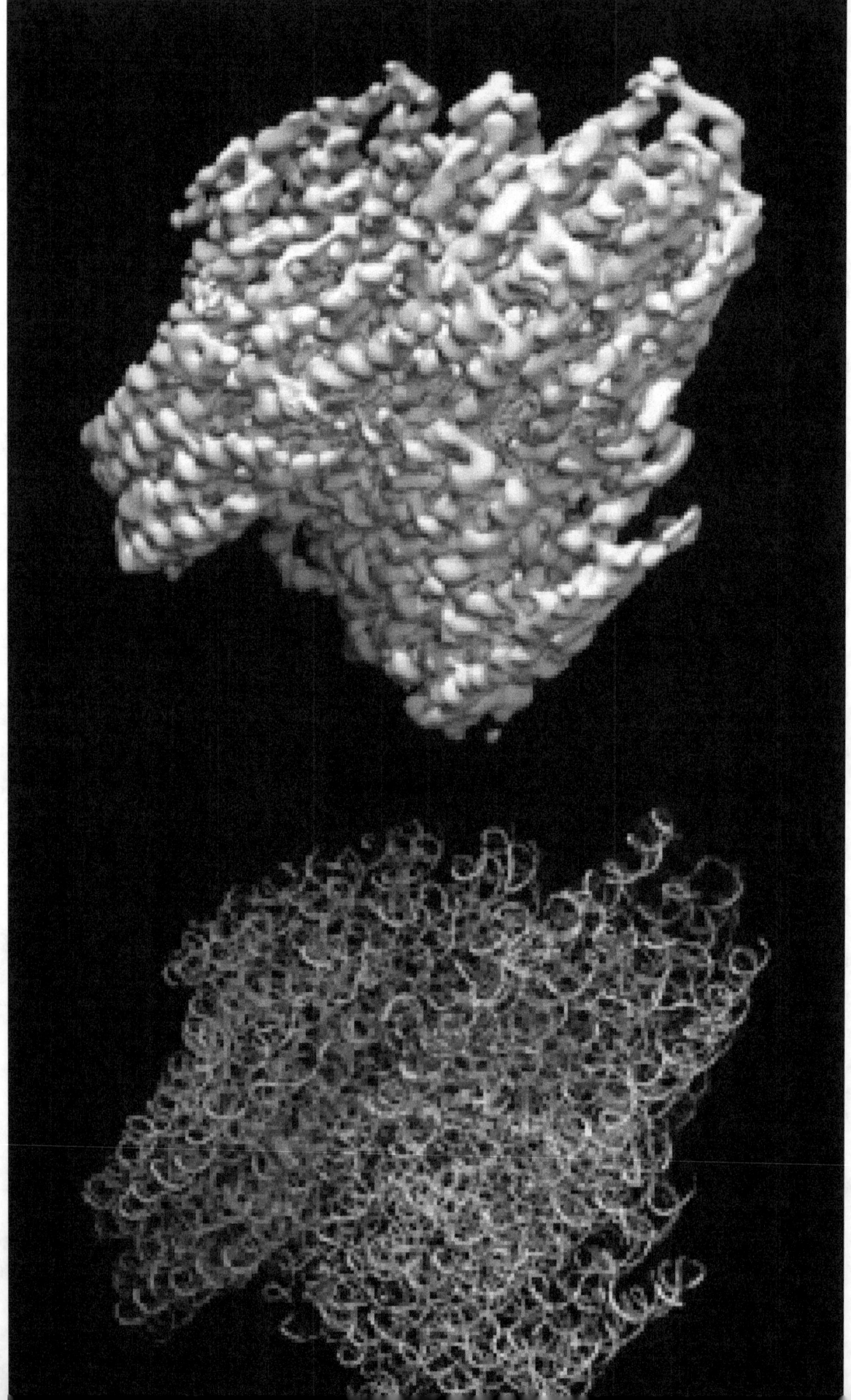

Chapter 7

Inorganic Chromosome Based in Silicon

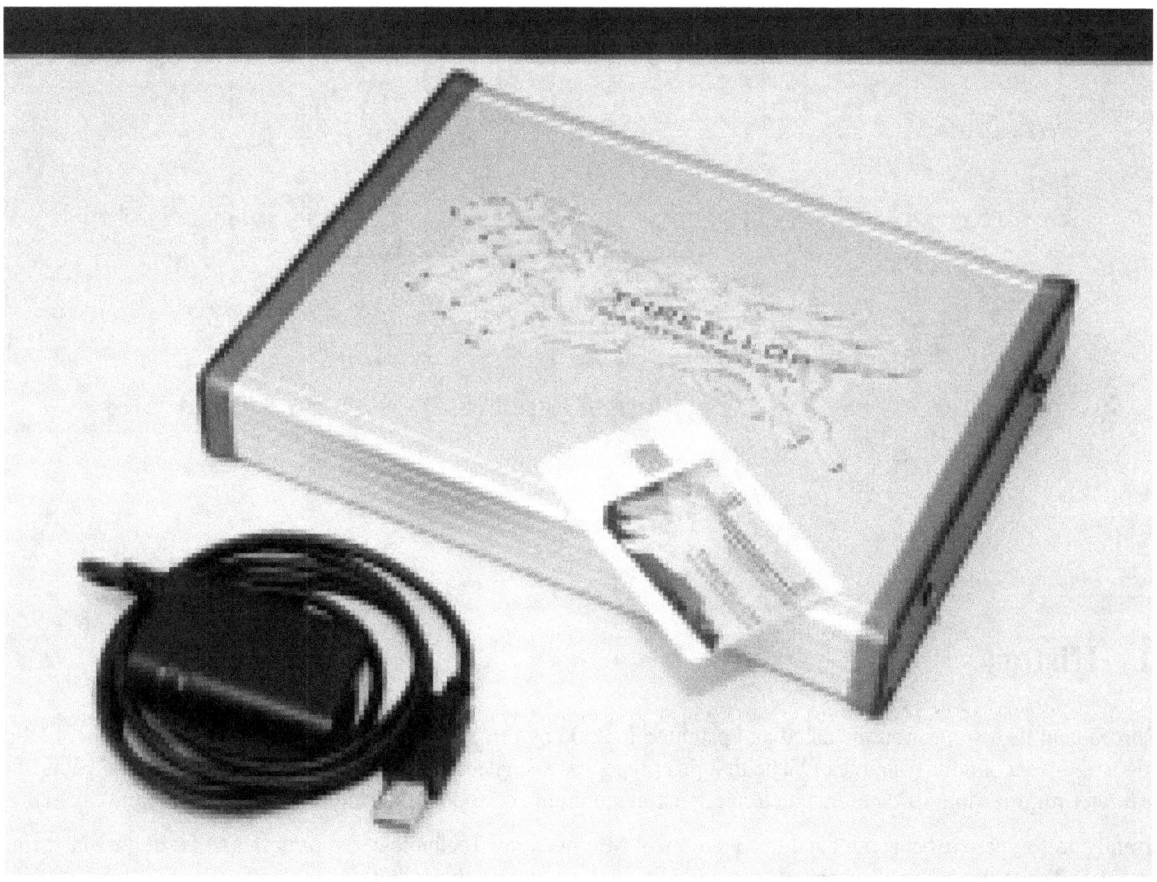

Inchrosil Device (Picture: J.Rayon)

Inorganic **Chro**mosome based in **Sil**icon **(InChroSil)** is an electronic device designed for storage and analysis of DNA sequences, with the organization of the fundamental memory units resembling the linear nature of DNA stands. This device was developed using the standard technology Complementary metal–oxide–semiconductor (CMOS).

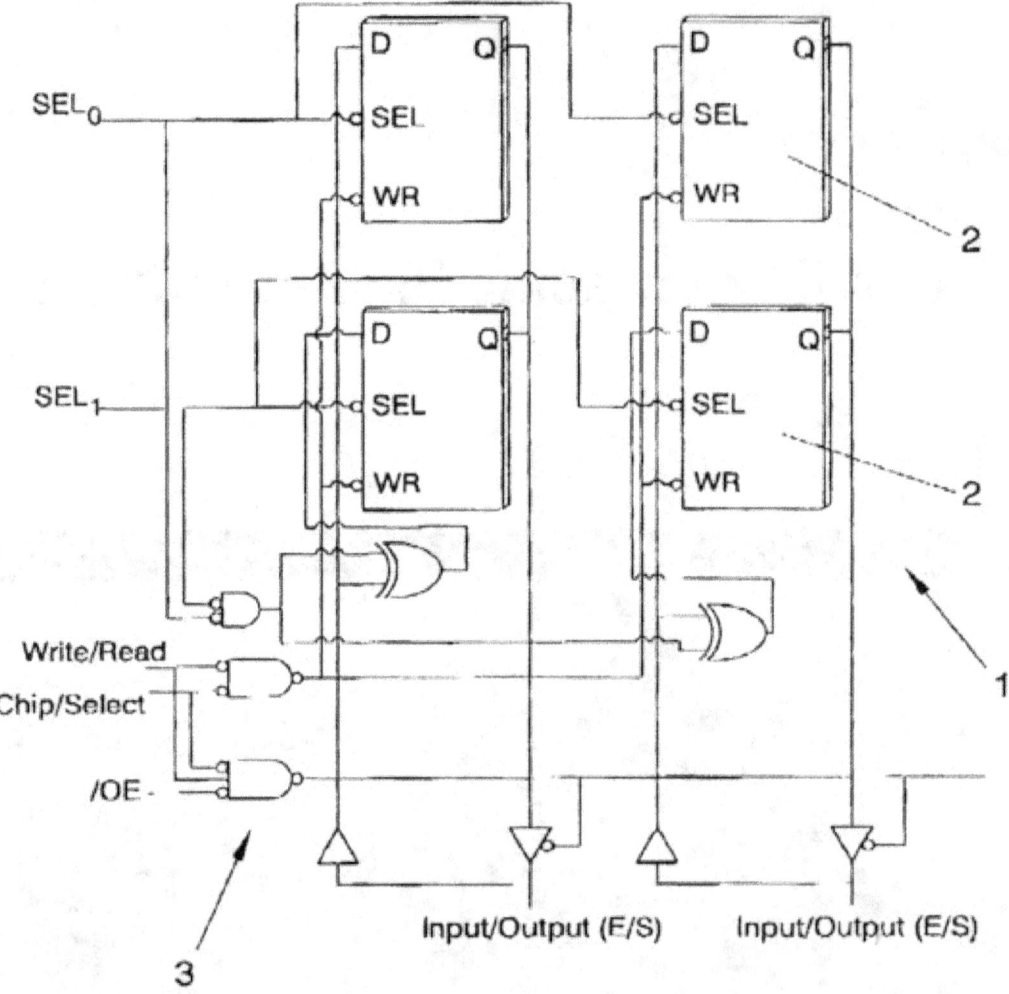

Figure 1 - Inchrosil's Unit

7.1 History

InChroSil and its systems were invented and patented in 2006 by three siblings, Silvia, Carlos and Jose Daniel Llopis at home with few economic resources.[1] The first prototype was a single nucleotide pair and was the size of a paper sheet. It was later improved into an integrated circuit in a configuration commonly referred to as a hybrid integrated circuit.

Currently prototypes are being built in the clean room of Microsystems Technology Laboratories **MTL** of the Massachusetts Institute of Technology.Threellop Nanotechnology Inc is the owner of the patent.

7.2 Inchrosil characteristics

InChroSil[2][3][4] permits the storage of genetic information in less space (88 percent savings), with higher performance [5][6] It can store the entire structure of the DNA (principal and secondary chains), meaning InChroSil can store and represent even incomplete chains including the existent holes within the DNA chain.

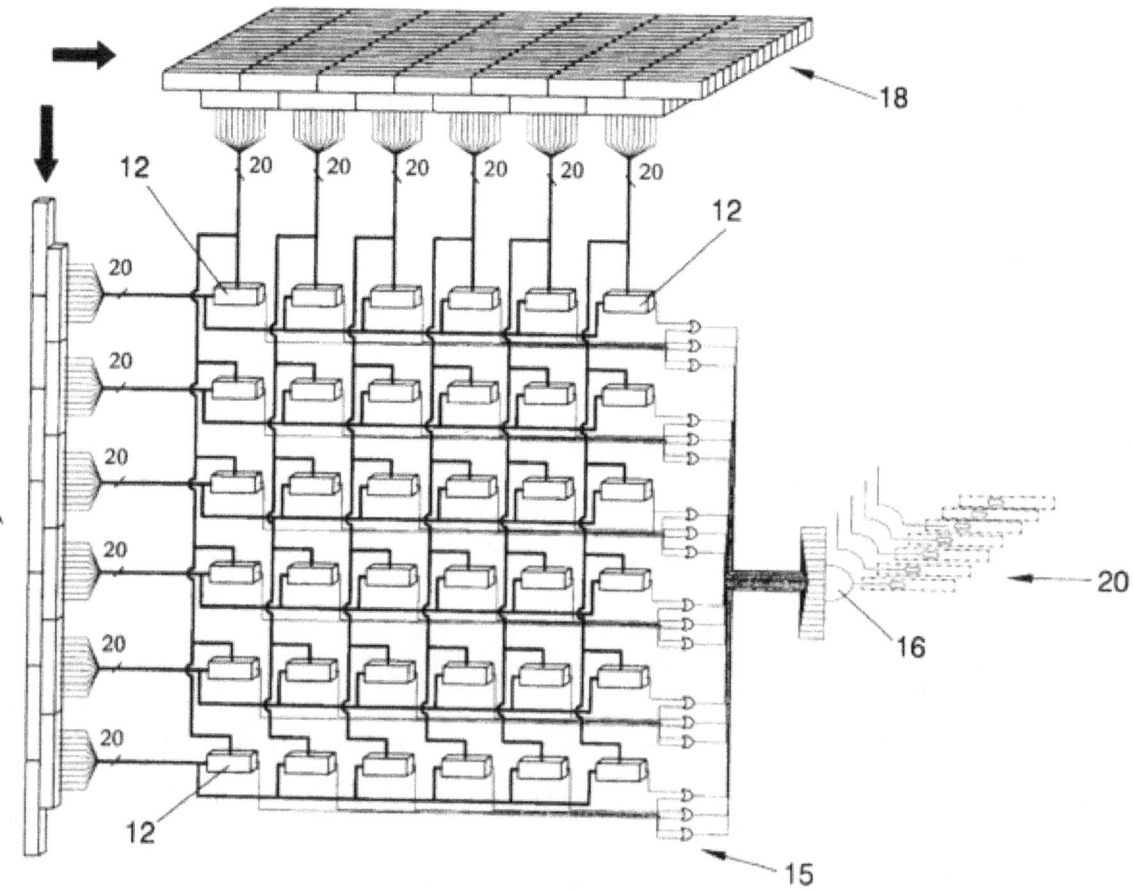

Figure 2 - Complete Circuit of Hamiltonian path problem

This device uses non-volatile rewritable digital memory and is also able to store non-genetic binary information in the nucleotides in a more compact form while retaining device, hardware and software compatibility.[7]

7.3   Adleman's experiment

Leonard Adleman's experiment computing a Hamiltonian path using DNA components,[8] was rebuilt by the Llopis siblings with inorganic electronic components. This system was intended to be less perishable than the organic materials and chemical reactions in industrial environments.

7.4   Uses for Inchrosil

Inchrosil is used mostly for mass storage of DNA sequences, with potential uses being:

- Fingerprinting staff,[9][10] with the storage of the genetic fingerprint. His invention is Dr. Alec Jeffreys at the

University of Leicester in 1984.[11]

- Genetic studies, with the tool where large sequences would be stored, which could be compared and manipulated digitally.

- Genebanks,[12] which store large amounts of genetic information.[13]

- Classification of species and animals.

- Tool medical studies (genetic).[14]

7.5   See also

- Nanotechnology

- Parallel computing

- DNA profiling

7.6   References

[1] Javier Carazo (2009-08-26). "Pequeños gigantes: La revolución española en el ADN". *Cinco dias.*

[2] Silvia, Carlos and Jose Daniel, llopis. "Electronic System to Emulate the Chain of the "DNA" Structure of a Chromosome". *PCT worldwide.* WIPO. Retrieved 7 November 2012.

[3] Hirsch, Cliff (November 2007). "Inchrosil - Inorganic Chromosome Based in Silicion". *Semiconductor Times.*

[4] Ministry of justice (Israel) (2009). "ELECTRONIC SYSTEM FOR EMULATING THE CHAIN OF THE DNA STRUC-TURE OF A CHROMOSOME". *Israel state scientific records.*

[5] de las heras (2009-09-27). "DNI genético, identificación segura". *Las provincias.*

[6] belt.es, El Portal de los Profesionales de la Seguridad. (September 2009). "Threellop Nanotechnology transformará el mundo de la salud, la seguridad y la defensa". *Nuevas Tecnologías Aplicadas a la Seguridad.*

[7] CDTI - Centro de desarrollo tecnologico e innovación, Governamment of Spain; Ministry of Industry (Spain) (July 2009). "Innovador sistema de Almacenamiento genético electrónico" (PDF). *Prespectiva* **31**: 60.

[8] Leonard M. Adleman. "Molecular Computation of solutions to combinational problems". Science, New Series, Vol. 266, No 5187, (Nov 11, 1994), 1021-1024.

[9] NanoSapin.org, Spanish Nanotechnology Network; Phantoms foundations (2009). "Threellop: Inorganic Chromosome based in Silicion". *Nanoscience and nanotechnology in Spain*: 168.

[10] University of Granada, Mando de Adiestramiento y Doctrina (MADOC); El Instituto Español de Estudios Estratégicos (11/12/2008). "Identification system by means of InChroSil (Inorganic Chromosome Based in Silicon)". *III Congreso Internacional de Seguridad y Defensa.* n III. Check date values in: |date= (help)

[11] Jeffreys A.J., Wilson V., Thein S.W. (1984). "Hypervariable 'minisatellite' regions in human DNA". Nature 314: 67–73. doi:10.1038/314067a0.

[12] ANTONIO GONZÁLEZ (11/06/2007). "Tarjetas de ADN para identificar soldados". *El publico.* Check date values in: |date= (help)

[13] Sanchis, Eva (2009-09-20). "Ya es posible comparar el ADN de forma masiva". *La Razon - Innovación.*

[14] Generalitat Valenciana, Government of Spain (2008). *Capacidades en Biotecnología - VIT SALUD.* Valencia: Generalitat Valenciana. p. 143.

7.7   External links

- Patent (WO/2009/022024) ELECTRONIC SYSTEM FOR EMULATING THE CHAIN OF THE DNA STRUC-TURE OF A CHROMOSOME

- European Patent (EP2180434A1) ELECTRONIC SYSTEM FOR EMULATING THE CHAIN OF THE DNA STRUCTURE OF A CHROMOSOME

- US Patent (US2010/0268520) ELECTRONIC SYSTEM FOR EMULATING THE CHAIN OF THE DNA STRUC-TURE OF A CHROMOSOME

Chapter 8

Thomas LaBean

Thomas (Thom) Henry LaBean is an American biochemist, bioengineer and associate professor at North Carolina State University. He was previously a research professor at Duke University. He is a leading researcher in the field of DNA nanotechnology.

LaBean graduated from Michigan State University in 1985 with a BSc in Biochemistry. In 1993 he was awarded a PhD in Biochemistry by the University of Pennsylvania.

8.1 Works

- *DNA-Templated Self-Assembly of Protein Arrays and Highly Conductive Nanowires*, Hao Yan, Sung Ha Park, Gleb Finkelstein, John H. Reif, Thomas H. LaBean, Science, 26, 2003, 1882-1884

- *Nanofabrication by DNA self-assembly (Review article)*, Hanying Li, Joshua D. Carter, Thomas H. LaBean, Materials Today, 2009

8.2 External links

- "Thomas H LaBean", *Scientific Commons*

- "Thomas H LaBean", *NCSU home page*

Chapter 9

List of DNA nanotechnology research groups

This list of **DNA nanotechnology research groups** gives a partial overview of academic research organisations in the field of DNA nanotechnology, sorted geographically. Any sufficiently notable research group (which in general can be considered as any group having published in well regarded, high impact factor journals) should be listed here, along with a brief description of their research.

9.1 North America

9.2 Asia

9.3 Europe

9.4 References

[1] Seeman's Lab

[2] Shih's Lab

[3] Peng Yin's Lab

[4] DNA and Natural Algorithms Group

[5] The Pierce Lab

[6] The Ellington Lab

[7] The Sleiman Lab

[8] Vallée-Bélisle's Lab's

[9] Rondelez's group

[10] Center for DNA nanotechnology

[11] Sondes de Nanotubes de Carbone et Nano-Biotechnologie

[12] Biomolecular systems and bionanotechnology

[13] Bionanotechnology

[14] Self-assembled structures and devices

[15]

Chapter 10

Nucleic acid design

Nucleic acid design is the process of generating a set of nucleic acid base sequences that will associate into a desired conformation. Nucleic acid design is central to the fields of DNA nanotechnology and DNA computing.[2] It is necessary because there are many possible sequences of nucleic acid strands that will fold into a given secondary structure, but many of these sequences will have undesired additional interactions which must be avoided. In addition, there are many tertiary structure considerations which affect the choice of a secondary structure for a given design.[3][4]

Nucleic acid design has similar goals to protein design: in both, the sequence of monomers is rationally designed to favor the desired folded or associated structure and to disfavor alternate structures. However, nucleic acid design has the advantage of being a much computationally simpler problem, since the simplicity of Watson-Crick base pairing rules leads to simple heuristic methods which yield experimentally robust designs. Computational models for protein folding require tertiary structure information whereas nucleic acid design can operate largely on the level of secondary structure. However, nucleic acid structures are less versatile than proteins in their functionality.[2][5]

Nucleic acid design can be considered the inverse of nucleic acid structure prediction. In structure prediction, the structure is determined from a known sequence, while in nucleic acid design, a sequence is generated which will form a desired structure.[2]

10.1 Fundamental concepts

The structure of nucleic acids consists of a sequence of nucleotides. There are four types of nucleotides distinguished by which of the four nucleobases they contain: in DNA these are adenine (A), cytosine (C), guanine (G), and thymine (T). Nucleic acids have the property that two molecules will bind to each other to form a double helix only if the two sequences are complementary, that is, they can form matching sequences of base pairs. Thus, in nucleic acids the sequence determines the pattern of binding and thus the overall structure.[5]

Nucleic acid design is the process by which, given a desired target structure or functionality, sequences are generated for nucleic acid strands which will self-assemble into that target structure. Nucleic acid design encompasses all levels of nucleic acid structure:

- Primary structure—the raw sequence of nucleobases of each of the component nucleic acid strands;

- Secondary structure—the set of interactions between bases, i.e., which parts of which strands are bound to each other; and

- Tertiary structure—the locations of the atoms in three-dimensional space, taking into consideration geometrical and steric constraints.

One of the greatest concerns in nucleic acid design is ensuring that the target structure has the lowest free energy (i.e. is the most thermodynamically favorable) whereas misformed structures have higher values of free energy and are thus

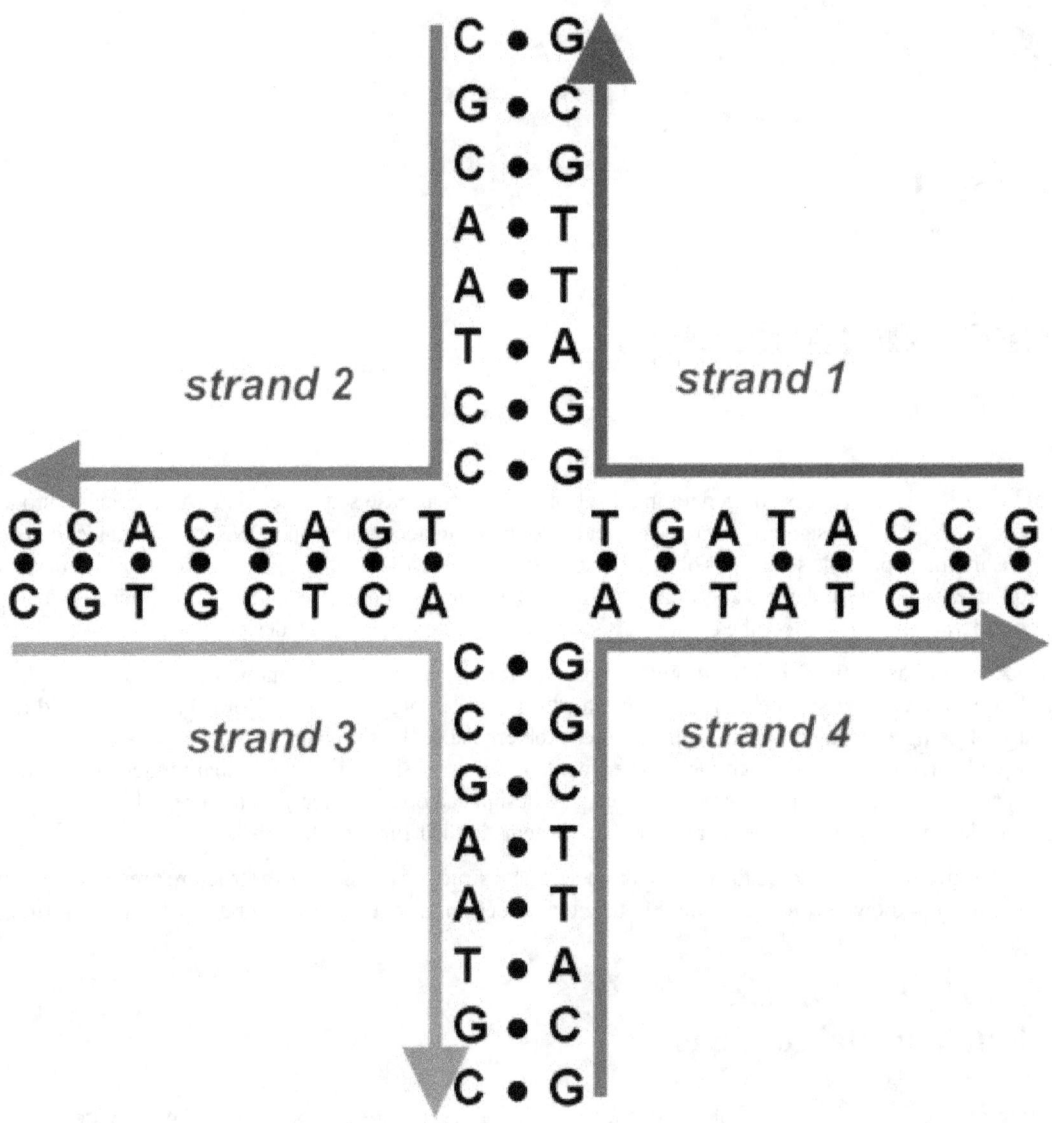

Nucleic acid design can be used to create nucleic acid complexes with complicated secondary structures such as this four-arm junction. These four strands associate into this structure because it maximizes the number of correct base pairs, with A's matched to T's and C's matched to G's. Image from Mao, 2004.[1]

unfavored.[2] These goals can be achieved through the use of a number of approaches, including heuristic, thermodynamic, and geometrical ones. Almost all nucleic acid design tasks are aided by computers, and a number of software packages are available for many of these tasks.

Two considerations in nucleic acid design are that desired hybridizations should have melting temperatures in a narrow range, and any spurious interactions should have very low melting temperatures (i.e. they should be very weak).[5] There is also a contrast between affinity-optimizing "positive design", seeks to minimize the energy of the desired structure in an absolute sense, and specificity-optimizing "negative design", which considers the energy of the target structure relative to those of undesired structures. Algorithms which implement both kinds of design tend to perform better than those that consider only one type.[2]

Chemical structure of DNA. Nucleic acid double helices will only form between two strands of complementary sequences, where the bases are matched into only A-T or G-C pairs.

10.2   Approaches

10.2.1   Heuristic methods

Heuristic methods use simple criteria which can be quickly evaluated to judge the suitability of different sequences for a given secondary structure. They have the advantage of being much less computationally expensive than the energy minimization algorithms needed for thermodynamic or geometrical modeling, and being easier to implement, but at the cost of being less rigorous than these models.

Sequence symmetry minimization is the oldest approach to nucleic acid design and was first used to design immobile versions of branched DNA structures. Sequence symmetry minimization divides the nucleic acid sequence into overlapping subsequences of a fixed length, called the criterion length. Each of the 4^N possible subsequences of length N is allowed to appear only once in the sequence. This ensures that no undesired hybridizations can occur which have a length greater than or equal to the criterion length.[2][3]

A related heuristic approach is to consider the "mismatch distance", meaning the number of positions in a certain frame where the bases are not complementary. A greater mismatch distance lessens the chance that a strong spurious interaction can happen.[5] This is related to the concept of Hamming distance in information theory. Another related but more involved approach is to use methods from coding theory to construct nucleic acid sequences with desired properties.

10.2.2 Thermodynamic models

Further information: Nucleic acid thermodynamics

Information about the secondary structure of a nucleic acid complex along with its sequence can be used to predict the thermodynamic properties of the complex.

When thermodynamic models are used in nucleic acid design, there are usually two considerations: desired hybridizations should have melting temperatures in a narrow range, and any spurious interactions should have very low melting temperatures (i.e. they should be very weak). The Gibbs free energy of a perfectly matched nucleic acid duplex can be predicted using a nearest neighbor model. This model considers only the interactions between a nucleotide and its nearest neighbors on the nucleic acid strand, by summing the free energy of each of the overlapping two-nucleotide subwords of the duplex. This is then corrected for self-complementary monomers and for GC-content. Once the free energy is known, the melting temperature of the duplex can be determined. GC-content alone can also be used to estimate the free energy and melting temperature of a nucleic acid duplex. This is less accurate but also much less computationally costly.[5]

Software for thermodynamic modeling of nucleic acids includes Nupack,[6][7] mfold/UNAFold,[8] and Vienna.[9]

A related approach, inverse secondary structure prediction, uses stochastic local search which improves a nucleic acid sequence by running a structure prediction algorithm and the modifying the sequence to eliminate unwanted features.[5]

10.2.3 Geometrical models

Geometrical models of nucleic acids are used to predict tertiary structure. This is important because designed nucleic acid complexes usually contain multiple junction points, which introduces geometric constraints to the system. These constraints stem from the basic structure of nucleic acids, mainly that the double helix formed by nucleic acid duplexes has a fixed helicity of about 10.4 base pairs per turn, and is relatively stiff. Because of these constraints, the nucleic acid complexes are sensitive to the relative orientation of the major and minor grooves at junction points. Geometrical modeling can detect strain stemming from misalignments in the structure, which can then be corrected by the designer.[4][11]

Geometric models of nucleic acids for DNA nanotechnology generally use reduced representations of the nucleic acid, because simulating every atom would be very computationally expensive for such large systems. Models with three pseudo-atoms per base pair, representing the two backbone sugars and the helix axis, have been reported to have a sufficient level of detail to predict experimental results.[11] However, models with five pseudo-atoms per base pair, explicitly including the backbone phosphates, are also used.[12]

Software for geometrical modeling of nucleic acids includes GIDEON,[11] Tiamat,[13] Nanoengineer-1, and UNIQUIMER 3D.[14] Geometrical concerns are especially of interest in the design of DNA origami, because the sequence is predetermined by the choice of scaffold strand. Software specifically for DNA origami design has been made, including caDNAno[15] and SARSE.[16]

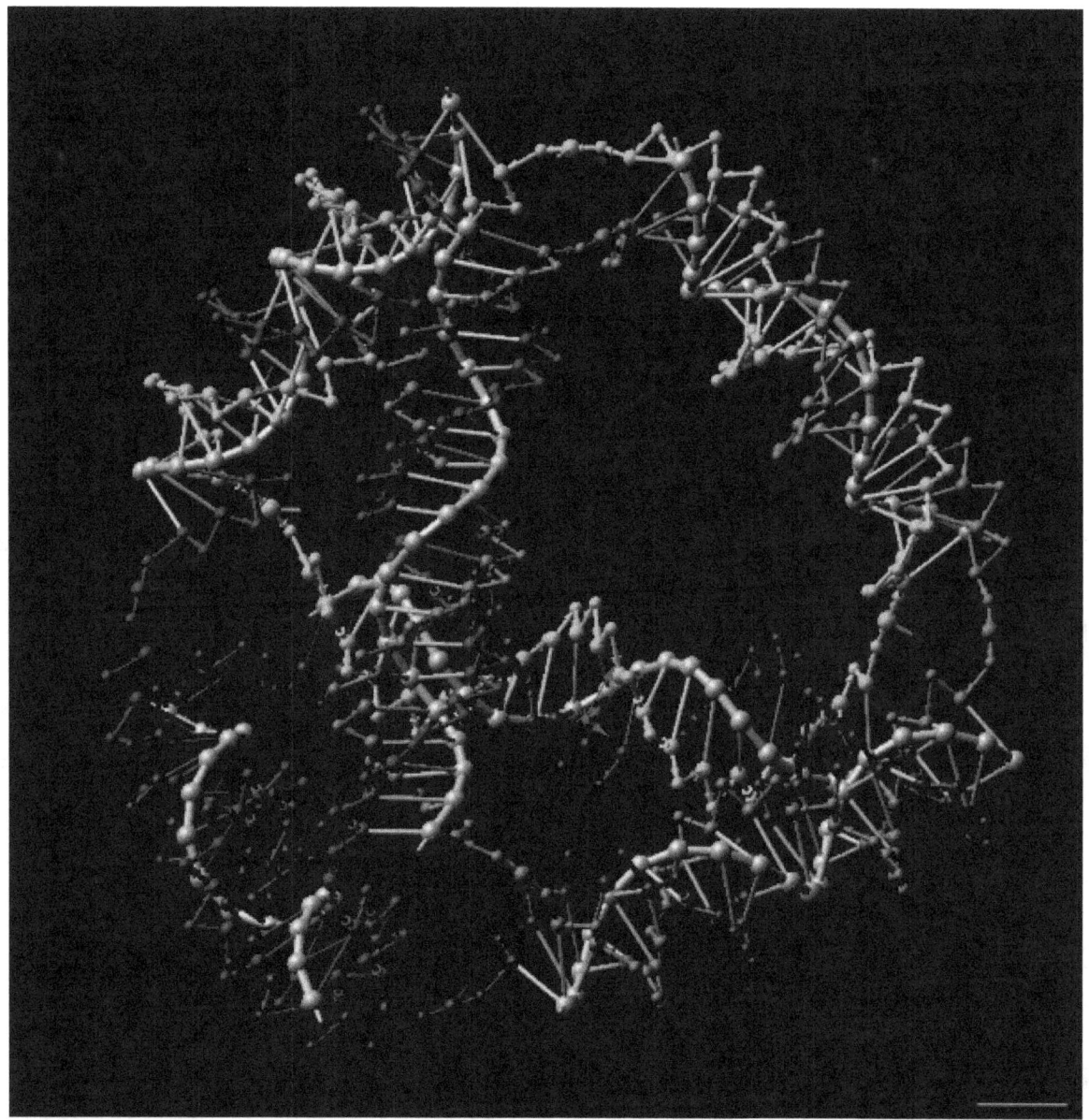

A geometrical model of a DNA tetrahedron described in Goodman, 2005.[10] Models of this type are useful for ensuring that tertiary structure constraints do not cause excessive strain to the molecule.

10.3   Applications

Nucleic acid design is used in DNA nanotechnology to design strands which will self-assemble into a desired target structure. These include examples such as DNA machines, periodic two- and three-dimensional lattices, polyhedra, and DNA origami.[2] It can also be used to create sets of nucleic acid strands which are "orthogonal", or non-interacting with each other, so as to minimize or eliminate spurious interactions. This is useful in DNA computing, as well as for molecular barcoding applications in chemical biology and biotechnology.[5]

10.4   See also

- Nucleic acid analogues

- Synthetic biology

10.5 References

[1] Mao, Chengde (December 2004). "The Emergence of Complexity: Lessons from DNA". *PLoS Biology* **2** (12): 2036–2038. doi:10.1371/journal.pbio.0020431. ISSN 1544-9173. PMC 535573. PMID 15597116.

[2] Dirks, Robert M.; Lin, Milo; Winfree, Erik & Pierce, Niles A. (2004). "Paradigms for computational nucleic acid design". *Nucleic Acids Research* **32** (4): 1392–1403. doi:10.1093/nar/gkh291. PMC 390280. PMID 14990744.

[3] Seeman, N (1982). "Nucleic acid junctions and lattices". *Journal of Theoretical Biology* **99** (2): 237–47. doi:10.1016/0022-5193(82)90002-9. PMID 6188926.

[4] Sherman, W; Seeman, N (2006). "Design of Minimally Strained Nucleic Acid Nanotubes". *Biophysical Journal* **90** (12): 4546–57. Bibcode:2006BpJ....90.4546S. doi:10.1529/biophysj.105.080390. PMC 1471877. PMID 16581842.

[5] Brenneman, Arwen; Condon, Anne (2002). "Strand design for biomolecular computation". *Theoretical Computer Science* **287**: 39. doi:10.1016/S0304-3975(02)00135-4.

[6] Dirks, Robert M.; Bois, Justin S.; Schaeffer, Joseph M.; Winfree, Erik; Pierce, Niles A. (2007). "Thermodynamic Analysis of Interacting Nucleic Acid Strands". *SIAM Review* **49**: 65. Bibcode:2007SIAMR..49...65D. doi:10.1137/060651100.

[7] Zadeh, Joseph N.; Wolfe, Brian R.; Pierce, Niles A. (2011). "Nucleic acid sequence design via efficient ensemble defect optimization". *Journal of Computational Chemistry* **32** (3): 439–452. doi:10.1002/jcc.21633. PMID 20717905.

[8] Zuker, M. (2003). "Mfold web server for nucleic acid folding and hybridization prediction". *Nucleic Acids Research* **31** (13): 3406–15. doi:10.1093/nar/gkg595. PMC 169194. PMID 12824337.

[9] Gruber AR, Lorenz R, Bernhart SH, Neuböck R, Hofacker IL (2008). "The Vienna RNA websuite". *Nucleic Acids Res* **36** (Web Server issue): W70–4. doi:10.1093/nar/gkn188. PMC 2447809. PMID 18424795.

[10] Goodman, R.P.; Schaap, I.A.T.; Tardin, C.F.; Erben, C.M.; Berry, R.M.; Schmidt, C.F.; Turberfield, A.J. (9 December 2005). "Rapid chiral assembly of rigid DNA building blocks for molecular nanofabrication". *Science* **310** (5754): 1661–1665. Bibcode:2005Sci...310.1661G. doi:10.1126/science.1120367. ISSN 0036-8075. PMID 16339440.

[11] Birac, Jeffrey J.; Sherman, William B.; Kopatsch, Jens; Constantinou, Pamela E.; Seeman, Nadrian C. (2006). "Architecture with GIDEON, a program for design in structural DNA nanotechnology". *Journal of Molecular Graphics and Modelling* **25** (4): 470–80. doi:10.1016/j.jmgm.2006.03.005. PMC 3465968. PMID 16630733.

[12] "PAM3 and PAM5 Model Descriptions". *Nanoengineer-1 documentation wiki*. Nanorex. Retrieved 2010-04-15.

[13] Williams, Sean; Lund, Kyle; Lin, Chenxiang; Wonka, Peter; Lindsay, Stuart; Yan, Hao (2009). "Tiamat: A Three-Dimensional Editing Tool for Complex DNA Structures". *DNA Computing*. Lecture Notes in Computer Science **5347**. Springer Berlin / Heidelberg. pp. 90–101. doi:10.1007/978-3-642-03076-5_8. ISBN 978-3-642-03075-8. ISSN 0302-9743.

[14] Zhu, J.; Wei, B.; Yuan, Y.; Mi, Y. (2009). "UNIQUIMER 3D, a software system for structural DNA nanotechnology design, analysis and evaluation". *Nucleic Acids Research* **37** (7): 2164–75. doi:10.1093/nar/gkp005. PMC 2673411. PMID 19228709.

[15] Douglas, S. M.; Marblestone, A. H.; Teerapittayanon, S.; Vazquez, A.; Church, G. M.; Shih, W. M. (2009). "Rapid prototyping of 3D DNA-origami shapes with caDNAno". *Nucleic Acids Research* **37** (15): 5001–6. doi:10.1093/nar/gkp436. PMC 2731887. PMID 19531737.

[16] Andersen, Ebbe S.; Dong, Mingdong; Nielsen, Morten M.; Jahn, Kasper; Lind-Thomsen, Allan; Mamdouh, Wael; Gothelf, Kurt V.; Besenbacher, Flemming; Kjems, JøRgen (2008). "DNA Origami Design of Dolphin-Shaped Structures with Flexible Tails". *ACS Nano* **2** (6): 1213–8. doi:10.1021/nn800215j. PMID 19206339.

10.6 Further reading

- Brenneman, Arwen; Condon, Anne (2002). "Strand design for biomolecular computation". *Theoretical Computer Science* **287**: 39. doi:10.1016/S0304-3975(02)00135-4.—A review of approaches to nucleic acid primary structure design.

- Dirks, Robert M.; Lin, Milo; Winfree, Erik; Pierce, Niles A. (2004). "Paradigms for computational nucleic acid design". *Nucleic Acids Research* **32** (4): 1392–1403. doi:10.1093/nar/gkh291. PMC 390280. PMID 14990744.— A comparison and evaluation of a number of heuristic and thermodynamic methods for nucleic acid design.

- Seeman, N (1982). "Nucleic acid junctions and lattices". *Journal of Theoretical Biology* **99** (2): 237–47. doi:10.1016/0022-5193(82)90002-9. PMID 6188926.—One of the earliest papers on nucleic acid design, describing the use of sequence symmetry minimization to construct immoble branched junctions.

- Andersen, Ebbe Sloth (2010). "Prediction and design of DNA and RNA structures". *New Biotechnology* **27** (3): 184–193. doi:10.1016/j.nbt.2010.02.012. PMID 20193785.—A review comparing the capabilities of available nucleic acid design software.

Chapter 11

Niles Pierce

Niles A. Pierce is an American mathematician, bioengineer, and professor at the California Institute of Technology. He is a leading researcher in the fields of DNA computing and DNA nanotechnology. His research is focused on kinetically controlled DNA and RNA self-assembly. Pierce is working on applications in bioimaging.

Pierce graduated from Princeton University in 1993 with a BSE in Mechanical & Aerospace Engineering. He attended Oxford University as a Rhodes Scholar, completing a DPhil in Applied Mathematics in 1997. He joined the faculty of the California Institute of Technology in 2000.

11.1 Works

- *Next-generation in situ hybridization chain reaction: higher gain, lower cost, greater durability* H.M.T. Choi, V.A. Beck, and N.A. Pierce ACS Nano 8(5):4284-4294, 2014.

- *Conditional Dicer substrate formation via shape and sequence transduction with small conditional RNAs* L.M. Hochrein, M. Schwarzkopf, M. Shahgholi, P. Yin, and N.A. Pierce J Am Chem Soc 135(46):17322-17330, 2013.

- *Nucleic acid sequence design via efficient ensemble defect optimization* J.N. Zadeh, B.R. Wolfe, and N.A. Pierce J Comput Chem 32:439-452, 2011.

- *NUPACK: Analysis and design of nucleic acid systems* J.N. Zadeh, C.D. Steenberg, J.S. Bois, B.R. Wolfe, M.B. Pierce, A.R. Khan, R.M. Dirks, and N.A. Pierce J Comput Chem 32:170-173, 2011.

- *Programmable in situ amplification for multiplexed imaging of mRNA expression* H.M.T. Choi, J.Y. Chang, L.A. Trinh, J.E. Padilla, S.E. Fraser, and N.A. Pierce Nature Biotechnol 28:1208-1212, 2010.

- *Programming biomolecular self-assembly pathways* P. Yin, H.M.T. Choi, C.R. Calvert, and N.A. Pierce Nature 451:318-322, 2008.

- *Thermodynamic analysis of interacting nucleic acid strands* R.M. Dirks, J.S. Bois, J.M. Schaeffer, E. Winfree, and N.A. Pierce SIAM Rev 49(1):65-88, 2007.

- *Triggered amplification by hybridization chain reaction* R.M. Dirks and N.A. Pierce Proc Natl Acad Sci USA 101(43):15275-15278, 2004.

- *A synthetic DNA walker for molecular transport* J.-S. Shin and N.A. Pierce J Am Chem Soc 126:10834-10835, 2004.

11.2   Resources

- NUPACK is a growing software suite for the analysis and design of nucleic acid systems.

- Molecular Instruments is an academic resource dedicated to the development and support of programmable molecular technologies for reading out and regulating cell state.

11.3   External links

- "Niles A. Pierce", *Scientific Commons*

- "Niles A. Pierce", *Caltech home page*

Chapter 12

John Reif

For the Oklahoma Supreme Court justice, see John F. Reif.

John H. Reif (born 1951) is an American academic, and Professor of Computer Science at Duke University, who has made contributions to large number of fields in computer science: ranging from algorithms and computational complexity theory to robotics and to game theory.

12.1 Biography

John Reif received a B.S. (magna cum laude) from Tufts University in 1973, a M.S. from Harvard University in 1975 and a Ph.D. from Harvard University in 1977.[1]

From 1983 to 1986 he was Associate Professor of Harvard University, and since 1986 he has been Professor of Computer Science at Duke University. Currently he holds the Hollis Edens Distinguished Professor, Trinity College of Arts and Sciences, Duke University. Since 2011 he has also been Distinguished Adjunct Professor, Faculty of Computing and Information Technology (FCIT), King Abdulaziz University (KAU), Jeddah, Saudi Arabia.

John Reif is President of Eagle Eye Research, Inc.,[2] which specializes in defense applications of DNA biotechnology. He has also contributed to bringing together various disjoint research communities working in different areas of nano-sciences by organizing (as General Chairman) annual Conferences on "Foundations of Nanoscience: Self-assembled architectures and devices" (FNANO[3]) for last 10 years.

He has been awarded Fellow of the following organizations: American Association for the Advancement of Science, IEEE, ACM, and the Institute of Combinatorics.

He is the son of Arnold E. Reif.

12.2 Research contributions

John Reif has made contributions to large number of fields in computer science: ranging from algorithms and computational complexity theory to robotics and to game theory. He developed efficient randomized algorithms and parallel algorithms for a wide variety of graph, geometric, numeric, algebraic, and logical problems. His Google Scholar H-index[4] is 61.

In the area of robotics, he gave the first hardness proofs for robotic motion planning as well as efficient algorithms for a wide variety of motion planning problems.

He also has led applied research projects: parallel programming languages (Proteus System for parallel programming), parallel architectures (Blitzen, a massively parallel machine), data compression (massively parallel loss-less compression hardware), and optical computing (free-space holographic routing). His papers on these algorithmic topics can be downloaded here.

12.2.1 Research in nanoscience

More recently, he has centered his research in nanoscience and in particular DNA nanotechnology, DNA computing, and DNA nanorobotics. In the last dozen years his group at Duke has designed and experimentally demonstrated in the lab a variety of novel self-assembled DNA nanostructures and DNA lattices, including the first experimental demonstrations of molecular scale computation and patterning using DNA assembly. His group also experimentally demonstrated various molecular robotic devices composed of DNA, including one of the first autonomous unidirectional DNA walker that walked on a DNA track. He also has done significant work on controlling errors in self-assembly and the stochastic analysis of self-assembly.[5]

12.3 See also

- Kinodynamic planning

12.4 Publications

He is the author of over 200 publications.[6] A selection:

- 2003. Hao Yan, Thomas H. LaBean, Liping Feng, and John H. Reif, Directed Nucleation Assembly of Barcode Patterned DNA Lattices, Proceedings of the National Academy of Science(PNAS), Volume 100, No. 14, pp. 8103–8108 (July 8, 2003).

- 2004. Peng Yin, Hao Yan, Xiaoju G. Daniel, Andrew J. Turberfield, John H. Reif, A Unidirectional DNA Walker Moving Autonomously Along a Linear Track, Angewandte Chemie, Volume 43, Number 37, pp. 4906–4911 (Sept. 20, 2004).

- 2007. John H. Reif and Thomas H. LaBean, Autonomous Programmable Biomolecular Devices Using Self-Assembled DNA Nanostructures, Communications of the ACM (CACM), Volume 50, Issue 9, pp. 46–53 (Sept 2007).

- 2008. Peng Yin, Rizal F. Hariadi, Sudheer Sahu, Harry M.T.Choi, Sung Ha Park, Thomas H. LaBean, John H. Reif, Programming DNA Tube Circumferences, Science, Vol. 321. no. 5890, pp. 824–826, (August 8, 2008).

12.5 References

[1] Reif's Vita

[2] Eagle Eye Research, Inc.

[3] FNANO

[4] h-index

[5] His papers on these topics can be downloaded here.

[6] Publications:

  - Reif's publications organized by research area
  - Reif's publications chronographically ordered
  - Reif's publications listed on Duke Faculty Website
  - Reif's publications listed on Google Scholar Website

12.6 External links

- Reif's Personal Web page
- Reif's Duke Web page
- Project collaboration
- Reif's Family, Schooling, Work and Play

Chapter 13

Paul W. K. Rothemund

Paul Wilhelm Karl Rothemund is a senior research fellow at the Computation and Neural Systems department at Caltech.[1] He has become known in the fields of DNA nanotechnology and synthetic biology for his pioneering work with DNA origami. He shared both categories of the 2006 Feynman Prize in Nanotechnology with Erik Winfree for their work in creating DNA nanotubes, algorithmic molecular self-assembly of DNA tile structures, and their theoretical work on DNA computing.[2] Rothemund is also a 2007 recipient of the MacArthur Fellowship.[3]

13.1 Life

Rothemund graduated from Laconia High School, New Hampshire, in 1990. He was the team captain of the championship Laconia team for the television quiz show *Granite State Challenge*. After graduating, Rothemund studied as an undergraduate at Caltech from 1990–1994, where he was a resident of Ricketts House. He attained his Ph.D. from the University of Southern California in 2001.

As a research fellow at Caltech, Rothemund has developed a technique to manipulate and fold strands of DNA known as DNA origami. Eventually, Rothemund hopes that self-assembly techniques could be used to create a "programming language for molecules, just as we have programming languages for computers."[4] His work on large-scale sculptures of his DNA origami was exhibited at the Museum of Modern Art in New York from February 24 to May 12, 2008.[5]

13.2 References

[1] "Paul W.K. Rothemund official website". Retrieved 2007-12-28.

[2] "2006 Foresight Institute Feynman Prize". Foresight Institute. Retrieved 20 July 2012.

[3] "Paul Rothemund". MacArthur Foundation. 18 January 2007. Retrieved 20 July 2012.

[4] Pelesko, John A. (2007). *Self assembly: the science of things that put themselves together*. Boca Raton: Taylor & Francis. p. 259. ISBN 1584886870.

[5] Eischeid, John (11 April 2008). "When Art and Science Meet, Nanoscale Smiley Faces Abound". Scientific American. Retrieved 20 July 2012.

13.3 External links

- Paul Rothemund's home page

- March 2006 Nature paper on DNA origami

- March 2007 TED Talk

- September 2008 TED Talk

Chapter 14

Nadrian Seeman

Nadrian C. "Ned" Seeman (born December 16, 1945) is an American nanotechnologist and crystallographer known for inventing the field of DNA nanotechnology.[1][2]

Seeman studied biochemistry at the University of Chicago and crystallography at the University of Pittsburgh.[3] He became a faculty member at the State University of New York at Albany, and in 1988 moved to the Department of Chemistry at New York University.

He is most noted for his development of the concept of DNA nanotechnology beginning in the early 1980s.[1] In fall 1980, while at a campus pub, Seeman was inspired by the M. C. Escher woodcut *Depth* to realize that a three-dimensional lattice could be constructed from DNA. He realized that this could be used to orient target molecules, simplifying their crystallographic study by eliminating the difficult process of obtaining pure crystals.[4][5] In pursuit of this goal, Seeman's laboratory published the synthesis of the first three-dimensional nanoscale object, a cube made of DNA, in 1991. This work won the 1995 Feynman Prize in Nanotechnology.[6]

The concepts of DNA nanotechnology later found further applications in DNA computing,[7] DNA nanorobotics, and self-assembly of nanoelectronics.[8] He shared the Kavli Prize in Nanoscience 2010 with Donald Eigler "for their development of unprecedented methods to control matter on the nanoscale."[8][9] The goal of demonstrating designed three-dimensional DNA crystals was achieved by Seeman in 2009, nearly thirty years after his original elucidation of the idea.

14.1   Notable publications

- Seeman, N (1982). "Nucleic acid junctions and lattices". *Journal of Theoretical Biology* **99** (2): 237–47. doi:10.1016/0022-5193(82)90002-9. PMID 6188926.—Considered to be the earliest paper outlining the concepts of DNA nanotechnology[10]

- Chen, Junghuei; Seeman, Nadrian C. (1991). "Synthesis from DNA of a molecule with the connectivity of a cube". *Nature* **350** (6319): 631–3. doi:10.1038/350631a0. PMID 2017259.—The synthesis of the DNA cube

- Winfree, Erik; Liu, Furong; Wenzler, Lisa A.; Seeman, Nadrian C. (6 August 1998). "Design and self-assembly of two-dimensional DNA crystals". *Nature* **394** (6693): 529–544. doi:10.1038/28998. ISSN 0028-0836. PMID 9707114.—The synthesis of two-dimensional periodic lattices of double crossover molecules

- Mao, Chengde; Sun, Weiqiong; Shen, Zhiyong & Seeman, Nadrian C. (14 January 1999). "A DNA Nanomechanical Device Based on the B-Z Transition". *Nature* **397** (6715): 144–146. doi:10.1038/16437. ISSN 0028-0836. PMID 9923675. —The first DNA-based nanomechanical device

- Seeman, Nadrian C. (June 2004). "Nanotechnology and the double helix". *Scientific American* **290** (6): 64–75. doi:10.1038/scientificamerican0604-64. PMID 15195395.—A popular science article explaining the field of DNA nanotechnology

- Zheng, Jianping; Birktoft, Jens J.; Chen, Yi; Wang, Tong; Sha, Ruojie; Constantinou, Pamela E.; Ginell, Stephan L.; Mao, Chengde; Seeman, Nadrian C. (2009). "From molecular to macroscopic via the rational design of a self-assembled 3D DNA crystal". *Nature* **461** (7260): 74–7. doi:10.1038/nature08274. PMC 2764300. PMID 19727196.—The synthesis of three-dimensional periodic lattices of tensegrity triangle molecules

- Gu, Hongzhou; Chao, Jie; Xiao, Shou-Jun; Seeman, Nadrian C. (2010). "A proximity-based programmable DNA nanoscale assembly line". *Nature* **465** (7295): 202–5. doi:10.1038/nature09026. PMC 2872101. PMID 20463734.—A DNA-based molecular assembly line

14.2 See also

- History of DNA nanotechnology

14.3 References

[1] Pelesko, John A. (2007). *Self-assembly: the science of things that put themselves together*. New York: Chapman & Hall/CRC. pp. 201, 242, 259. ISBN 978-1-58488-687-7.

[2] Dennis Overbye (3 June 2010). "8 Scientists Share $3 Million in Prizes". *The New York Times*.

[3] "Ned's Biography". Nadrian Seeman lab.

[4] Seeman, Nadrian C. (June 2004). "Nanotechnology and the double helix". *Scientific American* **290** (6): 64–75. doi:10.1038/scientificamerican0604-64. PMID 15195395.

[5] See Nadrian Seeman's homepage, Current crystallization protocol for a statement of the problem, and Nadrian Seeman's homepage, DNA cages containing oriented guests for the proposed solution.

[6] "1995 Feynman Prize in Nanotechnology Awarded For Pioneering Synthesis of 3-D DNA Objects". The Foresight Institute. 30 November 1995.

[7] Ng, W. D.; Wong, C. K. B. (2007). "Self-Recognition of DNA: From Life Processes to DNA Computation". *Biophysical Reviews and Letters* **2** (2): 123–137. doi:10.1142/S1793048007000490. PMC 2905173. PMID 20640192.

[8] "NYU Chemist Seeman Wins Kavli Prize in Nanoscience". New York University. 3 June 2010.

[9] "Names of the 2010 Kavli Prize winners announced". The Kavli Prize.

[10] Pinheiro, A. V.; Han, D.; Shih, W. M.; Yan, H. (December 2011). "Challenges and opportunities for structural DNA nanotechnology". *Nature Nanotechnology* **6** (12): 763–772. doi:10.1038/nnano.2011.187. PMC 3334823. PMID 22056726.

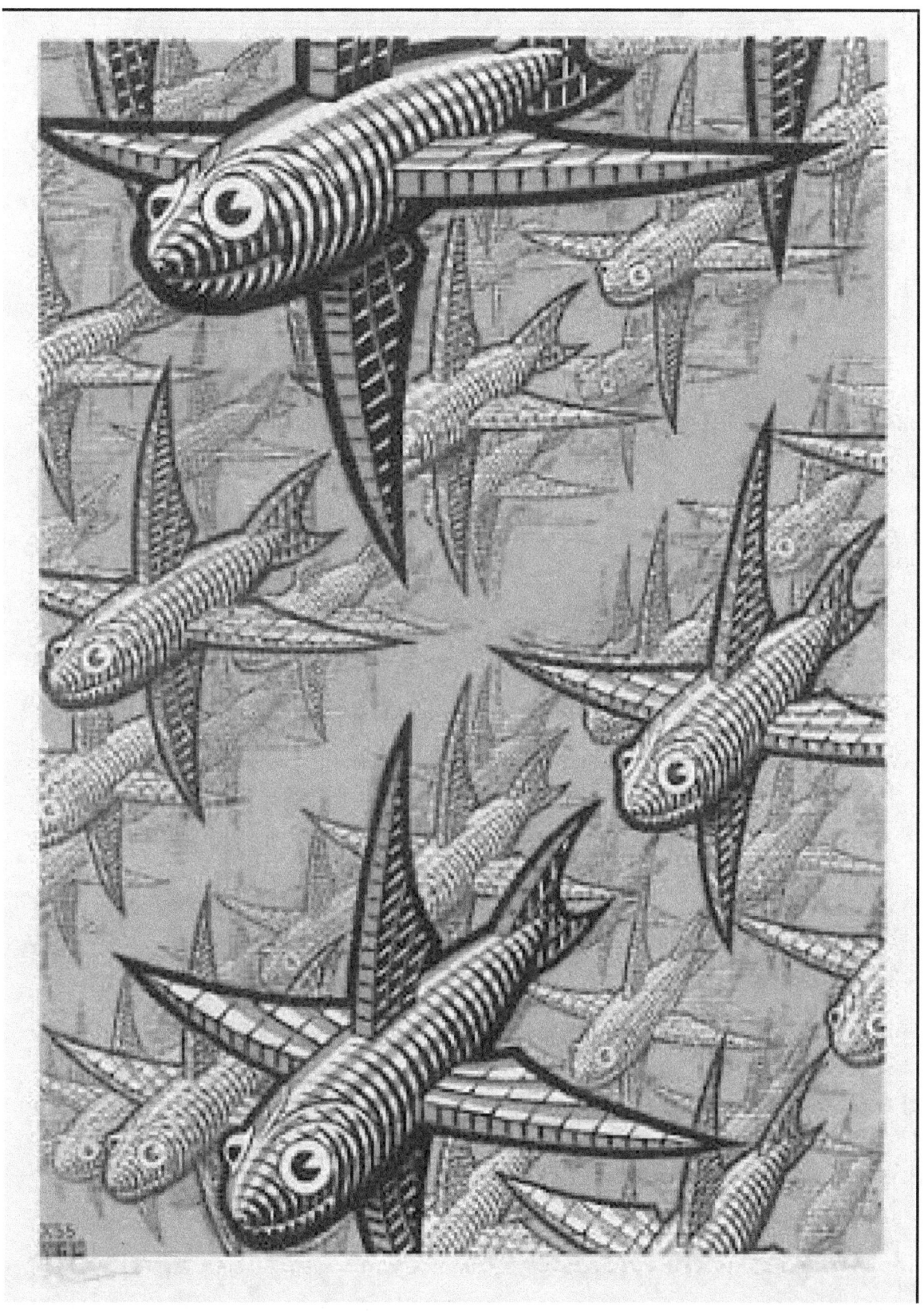

The M. C. Escher woodcut Depth *(pictured) inspired Nadrian Seeman to consider using three-dimensional lattices of DNA to orient hard-to-crystallize molecules. This led to the beginning of the field of DNA nanotechnology.*

Chapter 15

Friedrich Simmel

Friedrich C. Simmel (born 1970) is a German biophysicist and professor at the Technical University Munich. He is a researcher in the field of DNA nanotechnology and is best known for his work on DNA nanomachines[1] and dynamic DNA-based systems.[2]

Simmel received a PhD in experimental physics from the Ludwig Maximilian University of Munich in 1999. From 2000 to 2002 he was a PostDoctoral researcher at Bell Labs. He joined the faculty of the Technical University Munich as a full Professor in 2007.[3]

15.1 Awards and memberships

- 2006 Human frontier science program (HFSP) young investigator award

- 2009 Vice President (2009) of the International Society of Nanoscale Science, Computation and Engineering (ISNSCE)

- 2010 President (2009) of the International Society of Nanoscale Science, Computation and Engineering (ISNSCE)

- 2013 Elected Member of the National Academy of Science and Technology (acatech)

15.2 References

[1] Yurke, Bernard (2000), "A DNA-fuelled molecular machine made of DNA", *Nature* **406** (6796): 605, doi:10.1038/35020524, PMID 10949296.

[2] Franco, Elisa (2011), "Timing molecular motion and production with a synthetic transcriptional clock", *PNAS* **108** (40): E784, doi:10.1073/pnas.1100060108.

[3] *Prof. Dr. rer. nat. Friedrich Simmel*, Technische Universität München, retrieved 2014-11-13.

15.3 Works

- *DNA-based self-assembly of chiral plasmonic nanostructures with tailored optical response*, Anton Kuzyk, Robert Schreiber, Zhiyuan Fan, Günther Pardatscher, Eva-Maria Roller, Alexander Högele, Friedrich C. Simmel, Alexander O. Govorov, Tim Liedl, Nature, 483, 311-314, 2012

15.4   External links

- "Friedrich C. Simmel", *Scientific Commons*

- "Friedrich C. Simmel", *TU Munich home page*

- "Friedrich C. Simmel", *Article on a website run by the German Department of Education and Research (German language)*

Chapter 16

Transcriptor

For the first step in gene expression, see Transcription (genetics).

A **transcriptor** is a transistor-like device composed of DNA and RNA rather than a semiconducting material such as silicon. Prior to its invention in 2013, the transcriptor was considered the "final component required to build biological computers."[1]

16.1 Background

To function, a modern computer needs three different capabilities: It must be able to store information, transmit information between components, and possess a basic system of logic.[2] Prior to March 2013, scientists had successfully demonstrated the ability to store and transmit data using biological components made of proteins and DNA.[2] Simple two-terminal logic gates had been demonstrated, but required multiple layers of inputs and thus were impractical due to scaling difficulties.[3]

16.2 Invention and description

On March 28, 2013, a team of bioengineers from Stanford University led by Drew Endy announced that they had created the biological equivalent of a transistor, which they named a "transcriptor". That is, they created a three-terminal device with a logic system that can control other components.[2][3] The transcriptor regulates the flow of RNA polymerase across a strand of DNA using special combinations of enzymes to control movement.[1] According to project member Jerome Bonnet, "The choice of enzymes is important. We have been careful to select enzymes that function in bacteria, fungi, plants and animals, so that bio-computers can be engineered within a variety of organisms."[1]

Transcriptors can replicate traditional AND, OR, NOR, NAND, XOR, and XNOR gates with equivalents, which Endy dubbed "Boolean Integrase Logic (BIL) gates", in a single-layer process (i.e., without requiring multiple instances of the simpler gates to build up more complex ones).[2][3] Like a traditional transistor, a transcriptor can amplify an input signal.[1] A group of transcriptors can do almost any type of computing, including counting and comparison.[2][4]

16.3 Impact

Stanford dedicated the BIL gate's design to the public domain, which may speed its adoption.[1] According to Endy, other researchers were already using the gates to reprogram metabolism when the Stanford team published its research.[4]

Computing by transcriptor is still very slow; it can take a few hours between receiving an input signal and generating an output.[5] Endy doubted that biocomputers would ever be as fast as traditional computers, but added that is not the goal of his research. "We're building computers that will operate in a place where your cellphone isn't going to work", he said.[2] Medical devices with built-in biological computers could monitor, or even alter, cell behavior from inside a patient's body.[1] *ExtremeTech* writes:

> Moving forward, though, the potential for real biological computers is immense. We are essentially talking about fully-functional computers that can sense their surroundings, and then manipulate their host cells into doing just about anything. Biological computers might be used as an early-warning system for disease, or simply as a diagnostic tool ... Biological computers could tell their host cells to stop producing insulin, to pump out more adrenaline, to reproduce some healthy cells to combat disease, or to stop reproducing if cancer is detected. Biological computers will probably obviate the use of many pharmaceutical drugs.[1]

UC Berkeley biochemical engineer Jay Keasling said the transcriptor "clearly demonstrates the power of synthetic biology and could revolutionize how we compute in the future".[4]

16.4   References

[1] Sebastein Anthony (March 29, 2013). "Stanford creates biological transistors, the final step towards computers inside living cells". *Extreme Tech*. Retrieved March 29, 2013.

[2] Robert T. Gonzalez (March 29, 2013). "This new discovery will finally allow us to build biological computers". *IO9*. Retrieved March 29, 2013.

[3] Jerome Bonnet; Peter Yin; Monica E. Ortiz; Pakpoom Subsoontorn; Drew Endy (March 28, 2013). "Amplifying Genetic Logic Gates". *Science*. doi:10.1126/science.1232758.

[4] Lisa M. Krieger (March 29, 2013). "Biological computer created at Stanford". *San Jose Mercury News*. Retrieved March 29, 2013.

[5] Katherine Bourzac (March 28, 2013). "How to Make a Computer From a Living Cell". *MIT Technology Review* (Mashable). Retrieved March 30, 2013.

16.5   External links

- Jerome Bonnet; Peter Yin; Monica E. Ortiz; Pakpoom Subsoontorn; Drew Endy (March 28, 2013). "Amplifying Genetic Logic Gates". *Science*. doi:10.1126/science.1232758. - original journal article, published in *Science*

- Explanatory video created by Drew Endy

- NPR article with series of moving pictures that explain how the transcriptor works

- Public domain release of the BIL gates technology

Chapter 17

Andrew Turberfield

Andrew J Turberfield is a British Professor of Physics based at the University of Oxford. Turberfield's research is largely based on DNA nanostructures and photonic crystals,[1][2][3] and his work on both nanomachines and photonic crystals has been highly cited.[4] Turberfield is a fellow of Magdalen College, Oxford.[5]

17.1 References

[1] Jha, Alok (2000), "A flexible approach", *Nature* **408** (6812): 621, doi:10.1038/35046282, PMID 11117755.

[2] "The double helix becomes a pyramid", *New Scientist*, December 17, 2005.

[3] Turberfield. Andrew J Profile at the University of Oxford

[4] Web of Knowledge (requires institutional login) Publications ordered by citation

[5] Turberfield, Andrew Profile at Magdalen College

17.2 External links

- Turberfield DNA group at the University of Oxford

Chapter 18

Erik Winfree

Erik Winfree (born September 26, 1969[1]) is an American computer scientist, bioengineer, and professor at California Institute of Technology.[2] He is a leading researcher into DNA computing and DNA nanotechnology.[3][4][5]

In 1998, Winfree in collaboration with Nadrian Seeman published the creation of two-dimensional lattices of DNA tiles using the "double crossover" motif. These tile-based structures provided the capability to implement DNA computing, which was demonstrated by Winfree and Paul Rothemund in 2004, and for which they shared the 2006 Feynman Prize in Nanotechnology.[3][6]

In 1999, he was named to the MIT Technology Review TR100 as one of the top 100 innovators in the world under the age of 35.[7]

He graduated from the University of Chicago with a BS, and from the Computation and Neural Systems program at the California Institute of Technology with a PhD, where he studied with John Hopfield and Al Barr.[8] He was a Lewis Thomas Postdoctoral Fellow in Molecular Biology at Princeton University.[9] He was a 2000 MacArthur Fellow. His father Arthur Winfree, a theoretical biologist, was also a MacArthur Fellow.

18.1 Works

- *DNA Based Computers V: Dimacs Workshop DNA Based Computers V June 14–15, 1999 Massachusetts Institute of Technology*, Editors Erik Winfree, David K. Gifford, AMS Bookstore, 2000, ISBN 978-0-8218-2053-7

- *Evolution as computation: DIMACS workshop, Princeton, January 1999*, Editors Laura Faye Landweber, Erik Winfree, Springer, 2002, ISBN 978-3-540-66709-4

- "DNA Computing by Self-Assembly", *Ninth Annual Symposium on Frontiers of Engineering*, National Academies Press, 2004, ISBN 978-0-309-09139-8

- *Algorithmic Bioprocesses*, Editors Anne Condon, David Harel, Joost N. Kok, Arto Salomaa, Erik Winfree, Springer, 2009, ISBN 978-3-540-88868-0

18.2 References

[1] Erik Winfree resume

[2] Erik Winfree's homepage

[3] Pelesko, John A. (2007). *Self-assembly: the science of things that put themselves together.* New York: Chapman & Hall/CRC. pp. 201, 242, 259. ISBN 978-1-58488-687-7.

[4] "Biomolecular Computing" colloquium abstract

[5] Technology Review's 1999 TR35

[6] Seeman, Nadrian C. (June 2004). "Nanotechnology and the double helix". *Scientific American* **290** (6): 64–75. doi:10.1038/scientificamerican0
64. PMID 15195395.

[7] "1999 Young Innovators Under 35". Technology Review. 1999. Retrieved August 15, 2011.

[8] Erik Winfree's bio at Caltech Department of Computer Science

[9] Erik Winfree bio at Harvard

18.3 External links

- Erik Winfree at the Mathematics Genealogy Project
- "Erik Winfree", *Scientific Commons*

18.4 Text and image sources, contributors, and licenses

18.4.1 Text

- **DNA nanotechnology** *Source:* https://en.wikipedia.org/wiki/DNA_nanotechnology?oldid=686916149 *Contributors:* Shizhao, Giftlite, Stijn Ghesquiere, Thorwald, Iridia, Zlite, ZayZayEM, Anthony Appleyard, BD2412, Melesse, Rjwilmsi, Tony1, Kkmurray, Leptictidium, Nikkimaria, MaNeMeBasat, SmackBot, ShawnDouglas, Chris the speller, Epbr123, Gioto, Sangak, VoABot II, David Eppstein, Emw, Nanodetails, Schmloof, DrKay, Antony-22, Alnokta, Brvman, Izno, Malik Shabazz, Graham Beards, Cyfal, EoGuy, Niceguyedc, Piledhigheranddeeper, DragonBot, Aurelius173, Another Believer, Amaling, Dank, EEng, Addbot, Anthonydelaware, DOI bot, Baffle gab1978, Ettrig, AnomieBOT, 0x38I9J*, Kingpin13, Materialscientist, Citation bot, LilHelpa, P99am, Mouagip, Tolosthemagician, GrouchoBot, Pwkr, Mattimussi, Dcrjsr, Carlog3, Citation bot 1, Redrose64, Jonesey95, Specdude, Orenburg1, Trappist the monk, Jonkerz, RjwilmsiBot, Noommos, EmausBot, Beatnik8983, Slightsmile, Dcirovic, Hsleiman, ZéroBot, Bformhelix, Chalik1, Cberlind, Teapeat, ClueBot NG, Kirill Borisenko, Ankuzyk, Morgankevinj huggle, DeeperQA, Helpful Pixie Bot, Bibcode Bot, Razorbliss, Ginger Maine Coon, Aisteco, Justincheng12345-bot, Jicutler, Ling.Nut3, ChrisGualtieri, Lucquessoy, Khazar2, Dexbot, Br'er Rabbit, Leafonesky, JamesMadison Pres, Duckduckstop, Monkbot, Ittakesavillage2, Lilm345, Ashaul3 and Anonymous: 52

- **Coding theory approaches to nucleic acid design** *Source:* https://en.wikipedia.org/wiki/Coding_theory_approaches_to_nucleic_acid_design?oldid=663589226 *Contributors:* Lockley, MZMcBride, SmackBot, Chris the speller, JoergenB, Reedy Bot, Antony-22, VQuakr, Clausfse~enwiki, Yobot, Citation bot, LilHelpa, Jonesey95, RjwilmsiBot, Kushalsuryamohan, GoingBatty, Helpful Pixie Bot and Anonymous: 2

- **Robert Dirks** *Source:* https://en.wikipedia.org/wiki/Robert_Dirks?oldid=687398110 *Contributors:* Daniel Case, RussBot, Yoninah, Nikkimaria, DMacks, Ser Amantio di Nicolao, Victuallers, Antony-22, Yobot, Animalparty, Dexbot, Billdirks and Anonymous: 1

- **DNA computing** *Source:* https://en.wikipedia.org/wiki/DNA_computing?oldid=688801258 *Contributors:* Bdesham, Fred Bauder, Lexor, Wapcaplet, Ixfd64, Iluvcapra, SebastianHelm, Ahoerstemeier, Julesd, Pizza Puzzle, Dcoetzee, Arun.p.m, Wik, Atreyu42, Lowellian, P0lyglut, Henrygb, Giftlite, DavidCary, Drunkasian, Mintleaf~enwiki, Carlos-alberto-teixeira, Gdr, Zfr, Vivacissamamente, Rich Farmbrough, Michal Jurosz, JohnLynch, Tverbeek, Shanes, Ray Dassen, Maurreen, 4v4l0n42, Passw0rd, Arthena, Xeo~enwiki, Seans Potato Business, Suruena, TheoClarke, Kaster, Waldir, Lovro, Marudubshinki, Ketiltrout, Rjwilmsi, Kevmitch, Bgwhite, YurikBot, Vagodin, Piet Delport, Ihope127, Jsmaster24, WulfTheSaxon, Zwobot, Kkmurray, Rwwww, SmackBot, Rahulr7, Eaglizard, Edgar181, Powo, Ohnoitsjamie, Hmains, Alan smithee, Racklever, JonHarder, Memming, Radagast83, Wybot, Morio, Ligulembot, SashatoBot, MagnaMopus, Matt489, Xionbox, Aeternus, Rulesdoc, Lightofglory, Coconut99 99, Marek69, Alfalfahotshots, Gioto, Alphachimpbot, Pixelface, Wayiran, The Transhumanist, Dream Focus, Magioladitis, Pharod42, MartinBot, Theron110, Yaron K., Yonaa, Ceros, TimoSirainen, Maurice Carbonaro, Yonidebot, Antony-22, Jeff G., AlnoktaBOT, SCriBu, Medlakeguy, Luuva, Duncan.Hull, Billinghurst, Dirkbb, LoreMiles, Aruton, Denisarona, Francvs, Mr. Granger, ClueBot, PipepBot, Mild Bill Hiccup, DragonBot, Ykhwong, DumZiBoT, EastTN, Biomol, Dekart, Owl order, Addbot, DOI bot, Trak Nar, Luckas-bot, Yobot, AnomieBOT, Rubinbot, ThaddeusB, Jim1138, Sz-iwbot, Materialscientist, McBrayn, Citation bot, Xqbot, Sathimantha, Tolosthemagician, Stratocracy, FrescoBot, Dr. Megadeth, A little insignificant, DivineAlpha, Citation bot 1, Jonesey95, Rameshngbot, Mr Ape, Trappist the monk, LilyKitty, New-wuestenfux, RjwilmsiBot, EmausBot, John of Reading, Kushalsuryamohan, Wikipelli, Werieth, Aronlee90, Octavdruta, Dondervogel 2, Rucar, ChuispastonBot, Nanotube09, ClueBot NG, Horoporo, Helpful Pixie Bot, J.Dong820, Bibcode Bot, WarrenOutsky, Max Longint, Tomjones0001, Anubhab91, EricEnfermero, Dexbot, QuantumPhysicsPerson, Dustin V. S., Fixture, Monkbot, Sofia Koutsouveli, Verbal.noun, D.J.N.RauchmannStormannZnamenski-VienneLFV, Nøkkenbuer, Mahfuzur rahman shourov and Anonymous: 111

- **DNA machine** *Source:* https://en.wikipedia.org/wiki/DNA_machine?oldid=591036338 *Contributors:* Delirium, Robbot, Banus, NickelShoe, SmackBot, M stone, CmdrObot, Cydebot, WhaleyTim, Chaitanya.lala, Miniwheats, Antony-22, BuickCenturyDriver, Fadesga, TheRestIsNoise, Addbot, Abduallah mohammed, THEN WHO WAS PHONE?, Carlog3, LucienBOT, Dcirovic, Monkbot and Anonymous: 9

- **DNA origami** *Source:* https://en.wikipedia.org/wiki/DNA_origami?oldid=678177807 *Contributors:* Booyabazooka, Isopropyl, Jérôme, Woohookitty, David Haslam, Rjwilmsi, Bgwhite, Wavelength, Grafen, SmackBot, TestPilot, Delldot, Bluebot, Thumperward, Accurizer, Manoj 333, Alaibot, Nick Number, Midnightdreary, Nanodetails, Antony-22, Jgh32, Addbot, Dawynn, Bjornhogberg, Luckas-bot, Materialscientist, Citation bot, J04n, FrescoBot, Citation bot 1, Robfee, Trappist the monk, Jesus Presley, Jesse V., EmausBot, Dcirovic, ZéroBot, Mabster314, Chalik1, ClueBot NG, Latifahphysics, BG19bot, Ginger Maine Coon, Aisteco, Dexbot, Origamimonkey, Infiniti4, AmericanLemming, Kaygregory802, Beckarius, ATBlanchard and Anonymous: 18

- **Inorganic Chromosome Based in Silicon** *Source:* https://en.wikipedia.org/wiki/Inorganic_Chromosome_Based_in_Silicon?oldid=664801222 *Contributors:* Hooperbloob, Bgwhite, Cgingold, Antony-22, Dthomsen8, Yobot, Trappist the monk, MrX, EmausBot, Josve05a, Wbm1058, BattyBot, Viewoftech, Monkbot and OpenTechnology

- **Thomas LaBean** *Source:* https://en.wikipedia.org/wiki/Thomas_LaBean?oldid=482187550 *Contributors:* Edgar181 and Eracekat

- **List of DNA nanotechnology research groups** *Source:* https://en.wikipedia.org/wiki/List_of_DNA_nanotechnology_research_groups?oldid=629370132 *Contributors:* Welsh, Antony-22, MathewTownsend, Andre-LPN, Nedtech and Anonymous: 4

- **Nucleic acid design** *Source:* https://en.wikipedia.org/wiki/Nucleic_acid_design?oldid=667521097 *Contributors:* Daniel Case, Rjwilmsi, Wavelength, Nick Number, David Eppstein, Squidonius, Antony-22, Brvman, Addbot, DOI bot, Yobot, Citation bot, Trappist the monk, Dcirovic, Cberlind, Bibcode Bot, BG19bot, Anoyzz, Hamish59, Monkbot and Anonymous: 2

- **Niles Pierce** *Source:* https://en.wikipedia.org/wiki/Niles_Pierce?oldid=647742350 *Contributors:* Blanchardb, Helpful Pixie Bot, Eracekat, EditorMAK, ArmbrustBot and Anonymous: 1

- **John Reif** *Source:* https://en.wikipedia.org/wiki/John_Reif?oldid=650258708 *Contributors:* Michael Hardy, YUL89YYZ, Mdd, Woohookitty, Lockley, Stormbay, SmackBot, Neelix, Valrith, Waacstats, Fabrictramp, David Eppstein, GermanX, R'n'B, Johnpacklambert, Antony-22, Pharamond, Fnano~enwiki, JL-Bot, Msweir, Yobot, Johnreif, Richard.decal, FrescoBot, Tbhotch, RjwilmsiBot, Redmanhater, Extremepowskier69 and Anonymous: 7

18.4.2 Images

18.4.3 Content license

www.ingramcontent.com/pod-product-compliance
Lightning Source LLC
Chambersburg PA
CBHW080834180526
45168CB00006B/2680

* 9 7 8 1 5 1 9 1 1 6 2 6 0 *